Oral Communication Excellence for Engineers and Scientists:

Based on Executive Input

Synthesis Lectures on Professionalism and Career Advancement for Scientists and Engineers

Editors
Charles X Ling, *University of Western Ontario*
Qiang Yang, *Hong Kong University of Science and Technology*

Synthesis Lectures on Professionalism and Career Advancement for Scientists and Engineers includes short publications that will help students, young researchers, and faculty become successful in their research careers. Topics include those that help with career advancement, such as writing grant proposals; presenting papers at conferences and in journals; social networking and giving better presentations; securing a research grant and contract; starting a company, and getting a Masters or PhD degree. In addition, the series will publish lectures that help new researchers and administrators to do their jobs well, such as: how to teach and mentor, how to encourage gender diversity, and communication.

Oral Communication Excellence for Engineers and Scientists: Based on Executive Input
Judith Shaul Norback
July 2013

A Practical Guide to Gender Diversity for Computer Science Faculty
Diana Franklin
April 2013

A Handbook for Analytical Writing: Keys to Strategic Thinking
William E. Winner
March 2013

Oral Communication Excellence for Engineers and Scientists: Based on Executive Input
Judith Shaul Norback

ISBN: 978-3-031-01381-2 print
ISBN: 978-3-031-02509-9 ebook

DOI 10.1007/978-3-031-02509-9

A Publication in the Springer series
SYNTHESIS LECTURES ON PROFESSIONALISM AND CAREER ADVANCEMENT FOR SCIENTISTS AND ENGINEERS
Lecture #3

Series ISSN   1939-5221   Print   1939-523X Electronic

Oral Communication Excellence for Engineers and Scientists:

Based on Executive Input

Judith Shaul Norback
College of Engineering, Georgia Institute of Technology

*SYNTHESIS LECTURES ON PROFESSIONALISM AND CAREER
ADVANCEMENT FOR SCIENTISTS AND ENGINEERS #3*

ABSTRACT

Many of us have implemented oral communication instruction in our design courses, lab courses, and other courses where students give presentations. Others have students give presentations without instruction on how to become a better presenter. Many of us, then, could use a concise book that guides us on what instruction on oral communication should include, based on input from executives from different settings. This instruction will help our students get jobs and make them more likely to move up the career ladder, especially in these hard economic times.

Oral Communication Excellence for Engineers and Scientists: Based on Executive Input is the tool we need. It is based on input from over 75 executives with engineering or science degrees, leading organizations that employ engineers and scientists. For the presentation chapter, the executives described what makes a "stellar presentation." And for every other chapter, they gave input—on, for example, how to effectively communicate in meetings and in teams, how to excel at phone communication, how to communicate electronically to supplement oral communication, and how to meet the challenges of oral communication. They also provided tips on cross-cultural communication, listening, choosing the appropriate medium for a communication, elevator pitches, and posters; and using oral communication to network on the job.

Oral Communication Excellence for Engineers and Scientists includes exercises and activities for students and professionals, based on instruction that has improved Georgia Tech's students' presentation skills at a statistically significant level. Slides demonstrating best practices are included from Capstone Design students around the country.

KEYWORDS

workplace communication, engineering communication, science communication, workplace oral communication, engineering oral communication, science oral communication, workplace presentation, engineering presentation, science presentation, leadership communication, leadership oral communication, leadership skills, communication skills, industry communication, on-the-job communication, industry-related communication, executive communication, oral communication, executive input on communication, public speaking

Contents

Acknowledgements

A special thanks to Dr. Joel Sokol, Dr. Chen Zhou, Dr. Jane Ammons, Dr. Steve Hackman, Dr. Tristan Utschig (my research co-PI for seven years), Dr. Garlie Forehand, Dr. Susannah Howe, Dr. Elke Leeds. Thanks also goes to Dr. Franklin Bost, Dr. Ozlem Ergun, Dr. James Ferguson, Dr. Craig Forest, Dr. Xiaoming Huo, Dr. Pinar Keskinocak, Dr. Seong-Hee Kim, May Li, Dr. Jye-Chyi Lu, Dr. Yajun Mei, Eric Mungai, Dr. Wendy Newstetter, Dr. Gary Parker, Dr. James Rains, Lawrence Sharp, Dr. Ruth Streveler, Dr. Julie Swann, and Dr. Doug Williams. Appreciation for input goes to the many ISyE advisors of Capstone Design.

Thanks to the following executives: Mike Anderson, Leo Benatar, Bird Blitch, Larry Bradner, Pepper Bullock, Jim Butterworth, Jenifer Cistola, Douglas Coombs, Allan Dabbiere, Tom Daniel, Rich DeMillo, Sandy Dennis, Tom Dozier, Chuck Easley, Walt Ehmer, Ellen Ewing, Alan Fralick, Bill Gannon, Mel Hall, Dana Harmer, Betsy Higgins, Casey Hodgson, Ron Johnson, Louis Krost, Andrea Laliberte, Govantez Lowndes, Jody Markopoulos, Bob Martin, Hayne McCondichie, Andy McKenna, Debbie Meisel, Dennis Patterson, Don Pirkle, Shane Portfolio, Willis Potts, John Quinn, Paul Raines, Bill Reed, Brent Reid, Maria Rey, David Riviere, Chip Robert, Allen Robertson, Ed Rogers, Steve Rogers, Norberto Sanchez, Jane Snowdon, Phil Sokowitz, Jamie Striggs, Jeb Stewart.

Thanks to the following faculty and students who provided example Capstone Design slides: Dr. Brian Audiffred, Dr. Jim Baker, Dr. Franklin Bost, Dr. Nabila Bousaba, Dr. Larry Cartwright, Dr. Kevin Caves, Dr. Denny Davis, Dr. Rudy Eggert, Dr. William Endres, Dr. James Ferguson, Dr. Warren Hall, Dr. Joe Harris, Dr. David Koppelmen, Dr. David Mikesell, Dr. Ken Reid, Dr. Cameron Turner, Abhishek Anand, Stephanie Anderson, Sarah Armstrong, Leonora Baddoo, Jorge Baro, Katherine Baronowski, Chris Bauerlein, Dyanna Becker, Amy-Christina Biacobello, Mario Bojorquez, Adam Botts, Rebecca Bowden, Paul Brayford, Sarah Breen, Taylor Buono, Tyler Buran, Caitlyn Butler, Kari Caesar, Claire Choi, Gilbert Cha, Alexander Chretien, Callum Conaldson, Troy Connolly, Courtney Croft, Taylor Dalton, Warren Dantzler, Amanda Davis, Dustyn Deakins, Colin Doherty, Cody Dunwody, Christine Dzialo, Lauren Emory, Abuzer EmreOzer, Staci Feigenberg, Nathalie Flores, George Font IV, Allison Ford, Katy Gerecht, Chris Giardina, Sara Green, Patricia Groulx, Etta Grover-Silva, Kristyn Hall, Molly Harrington, Hollis Harris, Jennifer Harris, Justin Hill, Stephanie Hogge, Eric Holbrook, Reginald Holmes, Jacob Hora, Meghan Irving, Mark Jackson, Thaddeus Joseph, Hyunjin Jung, Khaled Kashlan, Matthew Kaufman, Nicholas Keller, Devon Kerns, Kanitha Kim, Kevin Kim, Benjamin Kingsdoft, Elizabeth Koenig, Rohit Kohli, Andrew Koo, Swetha Krishnakuman, William Landowski, Alan Lantz, Alexander Larsen, Karen

Li, Shaun Lim, Jordan Lolley, Sarah Mahon, Marni Mallari, Erika Mancha, James Matuszak, Paul McCrory, Myron McKeller, Brandon McLean, Doug Midkiff, Reem Mounsoura, Catherine Mulhern, Rohan Muthanna, Ida Ngambeki, Dean Nordhielm, Krysten Oates, Nnamdi Okafor, Joseph O'Meara, Leonard Oporto, Vishal Pachigar, Christopher Padgette, Allison Palsson, George Palumbo, Sangkeur Park, Kitu Patel, Dmitriv Pavlov, Johanna Pfeifer, Beth Pickett, Robert Pirkle, Mo Bill Qiao, Stan Radanov, Naveena Rahman, Rahul Reddy, Renee Redington, Kimberly Reinauer, Joaquin Reyes, Rahul Riddy, Jayt Roberts, Anna Robinson, Courtney Robinson, Julia Rodgers, Shane Rope, Daniel Routson, Avrigue Roy, Shane Saunders, Cloelle Sausville-Giddings, Tareq Shalhoub, Caitlyn Shea, Meghan Sheehy, Sam Shue, Courtney Sisson, Jennifer Sisson, Yvonne Smith, Jillian Spayde, Patrick Spence, Erica Spiritos, Pierce Spitler, Brad Stahl, John Steobar, David Streusand, Jared Stewart, Tom Stewart, Chloe Stoakes, Taronne Tabucchi, Meghan Taugher, Keith Taylor, Joseph Thaddeus, Rohan Thakur, Briana Tomboulian, Anh Tran, Peter Trocha, Jason Turner, Matthew Walker, David Walters, Eidan Webster, Jessica Wilbarger, Caitlin Wood, June Yeung, Shin Yang Yew, Colter Young, Joseph Zdon, Mimi Zhang, and Mohsin Zuberi.

Finally, thanks to Morgan & Claypool: Joel Claypool, Andrea Koprowicz, and Deb Gabriel, for making this book possible.

To Dick, always.

And to Garlie, mentor and friend extraordinaire.

With thanks to the Engineering Information Foundation, the National Science Foundation, the Alfred P. Sloan Foundation, and Georgia Tech ISyE alumni.

CHAPTER 1

Introduction

For years now, industry has expected better communication skills of engineers and scientists. Employers have been looking for oral communication excellence—the focus of this book. Written communication has received much more attention than oral communication, which has been studied in more detail only recently.

The problem of engineers and scientists lacking strong oral communication skills still remains today. Engineering students and practicing engineers generally cannot give stellar presentations or communicate flawlessly by phone, in meetings, and in teams. They have little notion about which medium (in-person, voicemail, E-mail, IM, texting) is most effective in particular situations. And students and professionals are not aware of how to use oral communication to network on the job.

Yet, engineering and science students and working engineers and scientists may think, since each of them communicates orally in many ways every day, how can they need more skill in oral communication? There may be parts of this book that some of these readers recognize. But this book covers so many areas of oral communication that, no matter how well the reader already communicates, all engineering and science students and professionals will find something useful here.

In our Workplace Communication Program at Georgia Tech, 170 students every semester make over 1,200 visits to the Workplace Communication Lab, where they receive instruction in oral communication. Since we opened in 2003 we have had over 19,500 student visits. Assessment has shown the instruction has improved the students' oral communication skills at a significant level (see References). Students have commented on their experience: they thought they were great presenters until they came into the Communication Lab. Then they learned how to improve. They highly recommend the Lab to future students and note particular skills they improved in. The students advise other students to be proactive, and they wonder how they can take the Communication Lab as a resource into the workplace with them.

Recent literature supports the need for a book like this. Starting with the Accreditation Board of Engineering and Technology (ABET) requirements in 2000, oral communication (and other skills) instruction for undergraduate engineers was necessary for an institution to remain accredited. Many studies in the last decade, including the landmark studies by the National Academy of Engineers (NAE; References) pointed out that the "Engineer of 2020" will need communication competencies including effective oral communication. Studies have shown that presentation skills are important and the demand for them continues to increase. But oral communication skills have not been studied much and there is a general lack of oral communication instruction in engineering courses. Educators disagree on exactly what communication skills are needed. A 2011 study at the Penn State Center for the Study of Higher Education (see reference list) followed up on the NAE study of the "Engineer of 2020" and found that faculty across institutions reported that communication is important in engineering curricula and there is some awareness of the NAE book's content.

This book responds to employers' demands for oral communication excellence in engineers and scientists. The book is unique because it is based on the input of executives with different engineering degrees working in a wide variety of settings. Successful businesspeople, or leaders, who started out with engineering degrees, provide input on the oral communication they expect of a new engineer or scientist and the skills needed for any engineer to quickly climb the career ladder. These guidelines are especially useful now, with the current economic situation in the U.S. Stellar oral communication skills will give engineering and science students and professional engineers and scientists an edge that will immediately show itself, in job interviews, co-ops, internships, and work on the job.

This book is also unique because it includes a trove of examples of situations from the real world. In the presentation chapter, for example, many slides from engineering capstone design courses involving real-world projects, in different disciplines and across the country, are provided. In other chapters, executives provide concrete examples of situations requiring excellent oral communication skills. For example, one executive described the actions of a newly hired engineer in a meeting they both attended. The executive had advised the new hire in advance not to contribute vocally to the meeting because of their lack of specific knowledge about the issues. Nevertheless, the new engineer did speak up—their comments implied the business group would take a very different direction than was currently expected. Their boss was forced to skillfully retract and undo the damage they had done. And of course it was a long time before the executive included the new hire in another meeting.

By reading this book and practicing with the exercises, you as the reader will improve your knowledge of how to use excellent oral communication skills in many key business situations.

This book begins with several chapters focusing on presentation. These chapters will help you develop the skill of giving creative and illustrative talks to varied audiences, with slides. You will also need to learn to present without slides, notes, or handouts. First you will find a chapter on background preparation and then four chapters relating to presentations: customizing to your audience, telling your story, displaying key information, and delivering your presentation. Many examples of slides created by engineering and science students in Capstone Design courses throughout the U.S. are used as examples as we discuss the 19 oral presentation skills identified by executives as key to giving a stellar presentation. Next you will see a chapter describing other oral communication skills, including:

- Common challenges in oral communication

- Choosing the right medium to support your oral communication

- Cross-cultural communication

- Listening

- Oral communication by phone

- Oral communication in meetings

- Oral communication in teams

- Using oral communication to build social networks on the job

Finally, you will see suggestions for advanced oral communication: giving a successful elevator pitch and preparing to present a poster.

CHAPTER 2

Background Preparation

2.1 INTRODUCTION

The National Academy of Engineering, the National Science Board, and various industry groups have stressed the importance of graduating engineers' and scientists' effective communication skills. Much past work has been done on researching and teaching engineering students writing skills, but only in the past decade has more emphasis been placed on the presentation skills of engineering students. Many studies have shown that communication skills are central to an engineer's and a scientist's success and promotion and that communication may account for a majority of the engineer's daily practice.

2.2 LEARNING ABOUT YOUR AUDIENCE

As we described above, customizing your presentation involves learning as much as possible about your audience before designing and presenting your slides. This information will enable you to tailor your presentation to their interests and needs, making your talk more interesting and engaging.

2.3 COLLECTING SPECIFIC KINDS OF AUDIENCE INFORMATION

Collect as much information as possible about the expected members of your audience. Answer the following questions through phone calls, E-mails, and the Web. The organizers sponsoring your talk will not mind your questions; instead they'll be impressed at your attention to detail.

2.3.1 HOW MANY PEOPLE WILL (OR COULD) BE IN YOUR AUDIENCE?

Do background research to avoid surprises. Five people definitely may be expected but more may come. The size of your audience and room will determine the volume of your presentation, but you also need to answer the following questions for each expected participant.

2.3.2 WHAT CONNECTION DOES EACH AUDIENCE MEMBER HAVE WITH YOUR PRESENTATION?

For example, are they:

- your main contact for the project you're presenting?

- the resource person who has provided you with relevant data?

- the executives of the company, with no interaction with your team before you give your presentation? If so, who keeps them updated about your project and how often?

- experts from another field? For example, if you are an industrial engineering student who is assisting a hospital with the layout of a proposed wing and if the architect joins your audience, they may have no prior connection to your work and may ask very different kinds of questions.

2.3.3 IS THE BACKGROUND OF YOUR AUDIENCE MEMBERS TECHNICAL, NON-TECHNICAL, OR BOTH?

Depending on the type of engineering student or engineer you are, you may present frequently to "mixed audiences" including engineers, managers who practiced engineering 10 years ago, and line workers with no engineering expertise. Or you may present to upper-level executives, a VP or a CEO, or to the heads of very different areas in a company.

When presenting to a mixed audience, you will need a different approach than you use when talking to a technical audience.

- Assume the lay audience members will not be familiar with technical terms.

- However, technical audience members may be irritated when you describe topics they already understand. Remind them—in a positive way—that not everyone in the audience knows what they know. One effective approach involves defining all the technical terms as you present, using introductory phrases such as, "To make sure we're all on the same page," or "We'll take a minute to review these definitions so we're all starting from the same point."

- Do not expect the lay part of your audience to remember new phrases. Insert key reminders here and there and spell out terms when possible instead of using acronyms or abbreviations.

Exercise 2.1. Write down several phrases or sentences you might use to orient your audience when you face a mixture of technical and non-technical people.

1.

2.

3.

Today and in the future, many practicing engineers work in multidisciplinary teams. For example, the team assembled to help fix Bridge 135 that collapsed in Minnesota included industrial, civil, and electrical engineers, financial experts, and managers. When one team member presented to the others, to be effective they avoided jargon and acronyms specific to their area. (Figures 2.1 and 2.2 present two slides showing terminology with pictures or diagrams.) For example, SKU or stock-keeping units, which describe the units stored in a warehouse, was familiar to industrial engineers but not necessarily the rest of the team. If you are a student, being a team member will better prepare you for the workforce. We will discuss acronyms in more depth in the topic of appropriate language later.

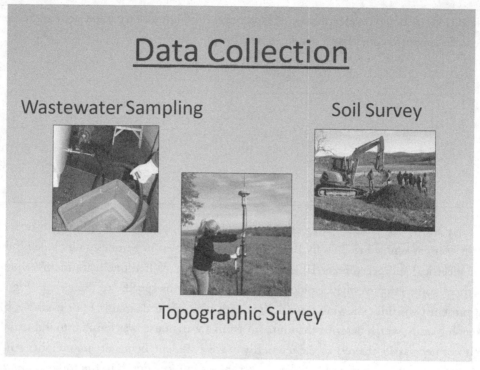

Figure 2.1: Environmental Engineering slide with terminology.

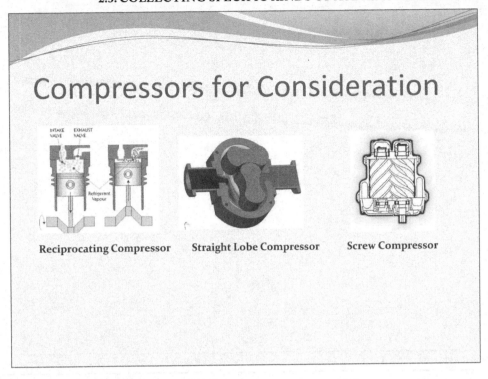

Figure 2.2: Mechanical Engineering with definitions.

Exercise 2.2. Write down three examples of specific terminology from your project—words or phrases that are unfamiliar to part of your audience. Then combine the terms with pictures or diagrams in the empty slide on the next page.

1.

2.

3.

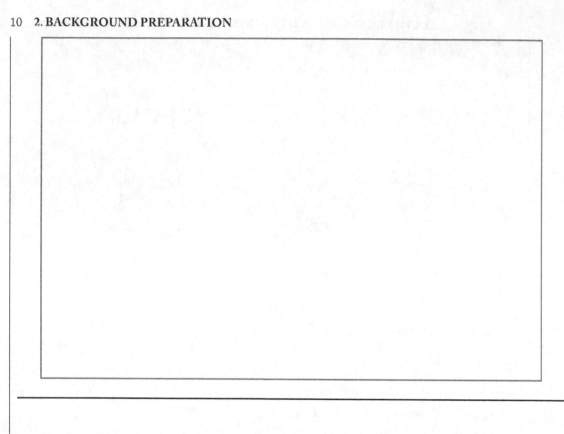

2.3.4 HOW MUCH DOES EACH AUDIENCE MEMBER KNOW ABOUT YOUR TOPIC? WHAT DO THEY UNDERSTAND THE PROBLEM TO BE?

What connection does each audience member have to the project or main topic of your presentation? To determine how much each audience member knows about your topic, search for related work on each audience member's website, ask your close contact, or call the person's assistant. Check web information first if possible so you don't waste your resource person's time. Group all your questions together and allow reasonable time for a response so you don't overburden the person you're calling.

Once you know the background knowledge each audience member has about your topic, you will know the range of information they have. This will help you orient your talk when you are brainstorming about questions likely to arise. You might get a detailed technical question from one audience member and straight-forward clarification questions (such as, "what does VNA mean?") from another. (VNA stands for "very narrow aisle.")

2.3.5   WHAT DOES EACH PERSON EXPECT YOU TO DO, AND WHAT DO THEY EXPECT TO GET OUT OF THE PRESENTATION?

While you learn about the topics described above, use the same methods to identify each individual's expectations for you and your team. For example, what steps do they expect you to take, and what recommendations are they expecting?

With this information you can tailor your talk by:

- describing only briefly what your audience expects, if they all agree;

- describing clearly and concisely the reasons for your next steps, if your audience members don't agree;

- dealing with disagreement, if it exists, or dealing with surprise if you are proposing a direction no one expects, by:

    ○ leaving enough time in your presentation for discussion of any disagreement;

    ○ laying the groundwork to avoid a surprise—by having discussions and gathering input from your audience members ahead of time.

If some of the audience members expect topics you hadn't expected to include, either clarify this by speaking to the appropriate people ahead of time, or add the topics to your talk as you present.

2.3.6   WHAT DOES EACH AUDIENCE MEMBER VALUE MOST?

Learning in advance what each member of your audience values most will help you decide the focus of your presentation. Many audience members will value the bottom line, in which case you need to explain, for example, whether the work will reduce the number of people in different departments or whether it will enable the individuals to become more efficient, thereby avoiding overtime. Other audience members may care more about market share or potential for new products, so you would focus on those topics.

2.3.7   WHAT IS THE RELATIVE AUTHORITY LEVEL OF EACH AUDIENCE MEMBER?

Before presenting, you will benefit by knowing the relative authority of each audience member: for example, who reports to whom, and who is or are the principle decision maker(s). During your talk, avoid looking only at the principle decision maker so the others will also feel respected. But you should check frequently with the highest-level audience members for non-verbal evidence of dissatisfaction or a question (a frown, for example.)

2.4   THE SETTING FOR YOUR PRESENTATION

Besides learning about your audience, learn ahead of time as much as possible about the setting for your talk. For example, visit the room or area you'll present in a day early, or at least an hour early. If you cannot visit the room, call ahead to ask about the room's size and seating and technology arrangement. The size of the room will tell you about how large a font you should use on your slides. The person sitting furthest away needs to be able to read everything. Remember, even if the room is medium-sized, one person may choose to sit in the last row.

Exercise 2.3 Describe the following aspects of the room for your next presentation.

1. Approximate size: _____

2. Number of chairs or seats: _____

3. Questions about layout:

    ° Where will the audience be? The seating may allow for a small number of people (including you) around a table, in which case you should check beforehand if your audience expects you to: 1) speak while seated from handouts you have in front of you; 2) speak while seated without handouts and with the actual slides projected on a screen in the room; or 3) speak while standing next to the slides which are projected on a screen. Answer:

    ° Where will you be standing? _____

    ° Where will the projector and screen be? _____

    ° Do the slides show up on the screen or does the room need to be darkened? What can you do to darken it? Close the blinds? Other?

4. Questions about handouts:

    ° What are your audience's expectations regarding when they receive copies of your slides? Your audience may have a strong preference for having the slides before them, since they will be able to make notes and may more easily follow your presentation. However, several executives have stressed to me that providing the audience with handouts before the presentation will distract them and therefore suggested giving out handouts at the end. For example, audience members may make notes on the slides or look ahead or back. Answer:

2.5   AVOIDING NOTES

Whatever the scenario is, never resort to notes, whether they are on handouts or your slide or whether they are on separate paper or cards.

If you use notes:

- the audience will view you as not bothering to learn the topics you're discussing.

- the audience will see you as unprofessional and disrespectful of them, and they will stop listening.

- you will have much less time to look at your audience. If you do make eye contact here and there, you will see evidence of "losing your audience."

A disengaged or uninterested audience shows its boredom by avoiding eye contact with you. If some members of the audience have stopped listening but are trying to be polite, you see blank stares. You may also hear papers rustling. Once the audience has stopped listening, you will have great difficulty in getting the audience members to "tune back in." You may try to engage the audience again by asking questions, speaking louder, and usimg a pause before and after your main points. You may refer directly to how your information meets your audience's specific needs. But if you are completely dependent on notes, you won't notice how the audience is reacting.

If you are presenting from slide handouts on the table in front of you, glance down only briefly to keep eye contact with your audience. In this situation your posture is important, since leaning forward with your hands on the table may seem aggressive to your audience.

When seated at a table but using slides projected onto a screen, use a laser pointer to help the audience follow the main points on the slides. Clearly, gestures you use while standing and presenting, such as pointing, are not as effective when you are seated. But using some hand movements to emphasize important points will help engage your audience, in contrast to keeping your hands below the table, or on the table, clasped.

2.5.1   USING A LASER SLIDE-ADVANCER

Practice before you present using the slide-advancer. Know exactly where to press to move the slides forward and back. If, for example, you show the prior slide instead of the next slide, you will distract your audience from your main message.

If you decide to use the pointer part of the slide-advancer, again, practice ahead of time. Most pointers are slightly different. You may not hit the right portion of the slide when you first practice, but you will soon get used to the pointer. Practice with it again in any new setting.

Use the pointer for very brief moments. For example, point to the box you're referring to and then turn off the pointer. This will help you avoid moving the pointer when you point, which distracts your audience from your message.

I once was in a presentation where the presenter had the pointer on for all of the presentation. I have also seen numerous presenters use their pointer to circle a key bit of information, four or five times. In each case even though I tried to listen to the speaker's message, it was hard. If a pointer is not available, refer verbally to where you are on the slide by saying, for example, "on the upper right part of the slide…"

CHAPTER 3

Presentation: Customizing to your Audience

We have discussed some of the implications of presenting while seated at a table. Presenters are often expected to stand near their slides while presenting, whether their audience is seated around an oval table, two tables in "V" format or in a large auditorium. One executive I spoke with said they preferred presenting in the "V" format, shown below.

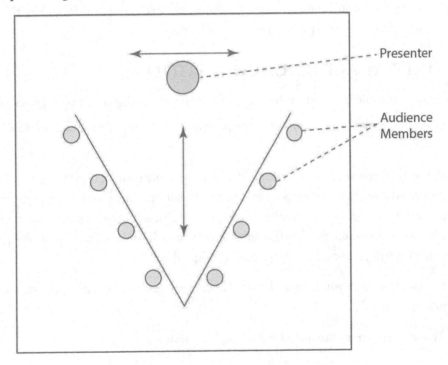

Figure 3.1: Example of presenting to an audience using a "V" seating arrangement.

As they showed their slides at the open end of the "V," they walked between the two tables, making eye contact with each audience member. This strategy kept their audience engaged.

If you are about to present in a large room or auditorium, remember the following:

• Even if only a few people come to your presentation, undoubtedly some will sit near the back.

- Make sure a person sitting in the back row can read all the information on your slides.

- Project your voice throughout so everyone can hear you. When you look for or ask for questions, check the people in the back of your audience as well as in front.

Ask about the technology setup and expectations before you present. For example, check where the technology will be, whether a microphone will be required and where it is located, whether you will be using a slide advancer and laser pointer, whether you need or want to bring your own laptop, how to test beforehand whether your software is compatible, and when you can test beforehand the size and color of the information on the screen.

Now that we have covered general information about customizing to your audience, we will discuss each skill separately. The skills are based directly on input from executives with engineering and science degrees who represent the consulting, manufacturing, banking, health care, venture capital, construction, home repair, and energy industries.

3.1   INITIAL AUDIENCE CONNECTION

Refer directly to audience needs to help define the purpose and goals of your presentation.

- Greet the audience as they enter the presentation setting, using each person's name if possible.

- When you start your presentation, refer directly to the context in which your audience is immersed to define the purpose and goals of your presentation. For example, if you are presenting a plan to increase the efficiency of the Emergency Department of a hospital, you would mention the barriers in that particular hospital currently resulting in long patient waiting times for assignments to hospital beds.

- Demonstrate that you have anticipated audience questions by having the relevant information at hand.

- Actively arrange to further address complex questions.

Exercise 3.1. Describe three things you can do to immediately connect to your audience.

1.

2.

3.

Have you used any recently? Yes___No___. If so, which ones have you used? Place a check by them.

3.2   USING APPROPRIATE LANGUAGE

Describe the concepts at just the right level for the audience. Specifically, you should:

1. clearly define and explain the use of technical terms,

2. use acronyms or abbreviations carefully, and

3. use appropriate grammar and spelling.

3.2.1   CLEARLY DEFINING AND EXPLAINING TECHNICAL TERMS

Explain important technical terms without additional complexity. The technical terms you use should be precise, not ambiguous. For example, the word "optimal" has a range of different meanings.

3.2.2   USING ACRONYMS OR ABBREVIATIONS CAREFULLY

If the audience has the **technical background** to understand the acronym, describe it when it's first used and then use the acronym through the rest of the presentation. For example, if you are an industrial engineering student or an industrial engineer presenting to other industrial engineers, use "SKU" with only a brief explanation to start since the audience understands this means "stock-keeping unit."

When you speak to engineers or scientists in other disciplines, assume you all share broader concepts, such as the use of the scientific method. When you present to others in your area, assume you share a great deal of abbreviations and acronyms.

If the audience is **non-technical**, use the whole phrase each time instead of the acronym or abbreviation. For example, if you are an industrial engineering student, you may be focusing on reducing the wait time in a hospital's emergency department, presenting to several nurse assistants, several nurses, and the head of the emergency department. You, as the presenter, need to do the extra work so the audience finds your talk easy to understand. If you use an unfamiliar acronym and only define it once before using it multiple times, your audience will need to do the work of remembering what the acronym stands for. This effort will distract your audience from your message.

Also, rephrase or simplify concepts as needed to suit the audience's lay background. For example, "Pareto analysis," an industrial engineering term, would not be understood by laypeople unless it is described as a statistical technique in decision-making.

Executives have mentioned that lay and technical audiences often have their own **company-specific acronyms,** in which case you would use those in your talk. For example, one company uses the word "apron" to refer to its employees on the floor who wait on the customers. Use the company-specific phrases or words to demonstrate your effort to customize the presentation to the audience. They will appreciate your professionalism.

Sometimes when speaking or practicing your talk, you may casually use some **informal terms,** such as "stuff" or "you guys." Always avoid this slang, replacing it with more professional language so the audience feels respected.

Exercise 3.2. List four acronyms or abbreviations unique to your field. Beside each, write your description for laypeople.

	Acronym or abbreviation	**Description for laypeople**
1		
2		
3		
4		

3.2.3   USING APPROPRIATE GRAMMAR AND SPELLING

Some people will say that even if a slide contains a "typo" (a typographical error) such as a misspelling, the main message still comes across. But many executives take exception to seeing even a single mistake. One executive of a health care firm told me that once they see a typo in a presentation, they stop listening. The single typo breaks the "trust" between the speaker and audience. After all, if a presenter is not meticulous enough to avoid a grammar or spelling error, how can the executive trust the other information presented? Data analysis and recommendations are suspect because of the "break in trust" caused by the typo. And, the executive added, once trust is broken it is very hard to rebuild. Many other executives expressed the same concern.

Check your slides for grammar and spelling errors by reviewing these guidelines, looking through your slide deck once for each possible error.

> Your capitalization should be consistent. For example, if you use initial capital letters on each of your slide titles and main headings, be consistent. Avoid having some main headings including only several words with initial capital letters. (Of course, some words, such as prepositions (to, from), are never capitalized.)

> Periods should be included on sentences but not on phrases. (As we will discuss later, more effective slides include only phrases, or a very small number of sentences and no phrases—often with a visual.)

> Check any demos for typos. Over the years I have seen many slides with typos, such as "abandonement" instead of "abandonment," "recomendations" instead of "recommendations," and a misspelling of a company name. In addition to breaking the trust between audience and presenter, typos often distract listeners from the main message. The audience may focus on the typo instead of listening to the speaker and following the slide.

3.3   RELEVANT DETAILS

The next part of customizing to the audience is using details relevant to the particular audience. Specifically, you should explain concepts very well. Use concrete examples audiences can really relate to. The details will vary with technical versus non-technical audiences. For example, if you are describing the processing in a rug factory, show the audience a variety of rug samples. And if you are talking in the abstract, end by giving a real-world example familiar to the audience.

3.4   TAKING QUESTIONS

Successful presenters adeptly accept and satisfactorily answer all audience questions. To prepare for questions, remember that many audience members, especially in the workforce, interrupt with questions rather than wait for the presentation to end. One executive from a consulting firm stressed that stellar presenters allow people to ask questions when they want, even if they seem to interrupt the presenter. Questions that interrupt are helpful to the communication between presenter and audience because the question arises in the part of the presentation that relates to the question.

For example, think of a presentation including a description of the client as part of the introduction, a description of approaches used, solutions, and recommended action. If an audience member has a question about the method used and waits to ask the question until after recommended actions, both presenter and questioner need to back up to the slides describing approaches used to establish the context of the question. But reversal of the logical flow can be avoided if audience members ask their questions during or right after the speaker presents the topic. One executive from the computer networking industry suggested that the presenter plan to stop after each major segment of their presentation and ask the audience if they have questions at that point.

Throughout the presentation, use eye contact to look for nonverbal signs from individuals. This topic is covered in more detail in the section on presentation delivery. You may see a frown (which may indicate confusion), a disgruntled look (which may represent disagreement), or a bored look (which may indicate inattention due to lack of understanding).

Before the presentation, brainstorm possible audience questions by thinking of the audience's background and interests while reviewing each slide. Presentations often have a time limit, so you may not be able to include all the answers to potential questions. But you should be prepared for the questions. In advance of the presentation, set aside additional information to have at hand to draw from when responding to questions. Include this more detailed information as slides in the appendix or possibly in a handout. To show you are prepared, demonstrate to your audience that certain questions have been anticipated. You may say,:

"Yes, we have thought about that question. . ."

or

"Since I expected that question to come up today, I've prepared an appendix to answer it."

This type of phrase is effective when used sparingly. Your preparation can also be shown nonverbally through your personal presence, for example, your eye contact and energy when answering the question. (Personal presence will be discussed in more detail later.) When taking questions you should show respect for the questioner, for example, even if they are a mechanic who is part of the audience in an industrial engineering student project.

Here are suggested steps for your success:

1. When an audience member raises their hand, complete your sentence before calling on them. You will be seen as more seasoned and professional. However, if an urgent question comes up, respond to it immediately.

2. Never speak over a person asking a question or cut in when you think you know what they are asking.

3. Listen carefully to the person's question to see if you understand it. If you are not sure, ask the audience member at this point for clarification or added information. Make sure you understand the question correctly before answering.

4. If you have a large audience, repeat the question and clarification more loudly so everyone can hear. Over the years I have seen very few audience members, in big audiences, who indicated they didn't hear a question, so the responsibility is yours as the presenter.

5. Answer the question briefly and to the point. If you don't know the answer, do not go on and on, hoping that something you eventually say will answer the question. This approach wastes everyone's time. Be sure to be honest and straightforward. One executive whose firm serves many different clients needing medical services stressed that when presenting to a potential client, he would sometimes say, "I don't know." He would then build trust with the client by making it a point to get them the information as soon as possible. You will need to determine for yourself just how comfortable you would be saying, "I don't know." Here are some other effective answers:

"We haven't looked into that yet, but we will before we report to you again."

"I'm not sure. Who should I talk with from your company about that?"

"I don't know but I'll look into that..."

"We haven't run that test yet; we'll make sure to keep that in mind when we continue our work..."

"I'm not sure how to answer that question. I'll check into it and have a response for you by ..." (and then remember to follow up)

6. Think of audience questions in a positive light instead of as a nuisance. Be open-minded. Treat the interaction with the audience as an information exchange. Often audience questions lead to better project or program results. And you will definitely demonstrate professionalism if you thank the audience member for their suggestions.

   As you finish answering the audience question, they may nod their head to indicate they understood your answer. If they don't, check with them verbally to be sure you answered their question. Simply saying "Is that clear?" or "Did I answer your question?" will determine if your communication has been successful. If you skip this step, misunderstandings may occur that can lead to major problems later on. If your questioner answers "no," then you need to clarify the question again and start over. If time is an issue, you may ask the audience member if you can meet with them after the talk.

7. If you are presenting with a group, when questions come up make sure only one group member answers the question. If, in contrast, one person answers the question and one or two others then add to the answer, the audience may see the second an-

swer as a correction of the first. As a colleague of mine says, "then you all look bad." The only time a second group member should speak up after a question is answered is when the first answer contains a very critical mistake. Here are ways to help avoid this embarrassing situation:

- Identify each group member's area of project expertise. Expect that person to answer questions arising in their area.

- When a group isn't sure which member can best answer a question, take a moment to make eye contact with each other. Then one person will "step forward" to take the question or slightly raise their index finger to indicate they will answer.

8. After you are sure you answered the question, transition back to the topic you were discussing. Add a little information to help audience members remember where you were.

Exercise 3.3. List three things you can do to improve how you handle audience questions. Identify the date when you will implement the changes.

1.   Date _____

2.   Date _____

3.   Date _____

CHAPTER 4

Presentation: Telling your Story

When preparing your slides and later giving your presentation, use a logical flow and make sure the different parts of the presentations are interconnected. As a result, you will have created a memorable unified message. Four skills or strategies you may use to produce this result are: sequencing, emphasizing key points, using appropriate context, and being sensitive to time. These are described below.

4.1 SEQUENCING

So your audience can easily follow your presentation, clearly link different parts of the presentation and use appropriate transitions. Do not speak on a subject without explaining it first, and use transitions as "signposts" to lead your audience through your presentation's flow. Group similar content in an easily followed order. Be sure the audience is never surprised by a topic coming "out of the blue." Instead lay the background before introducing new concepts.

4.1.1 STORYBOARDING

This approach is a simple yet remarkably effective way to ensure your logical flow. The practice is adopted from the film industry where a storyboard includes, for example, a string of pictures depicting small movements. When the pictures were stacked, and then run through very quickly, a larger movement appeared. (Think of your favorite old-time cartoon character. Every action or chase consisted of a multitude of separate pictures.)

When using storyboarding for presentations, start with a sheet with many equal-sized boxes in rows, as in Figure 4.1 on the following page. If your slides are already drafted, write the titles in each box, left to right, and top to bottom. If you are about to create your slides, write your expected topics in the boxes.

Make sure you include the following types of slides. They will be described in detail below.

- The cover slide

- The overview, describing what will be covered in the presentation

- The introduction/background slide

- The main body

- The summary or conclusions

- The "Next Steps" slide

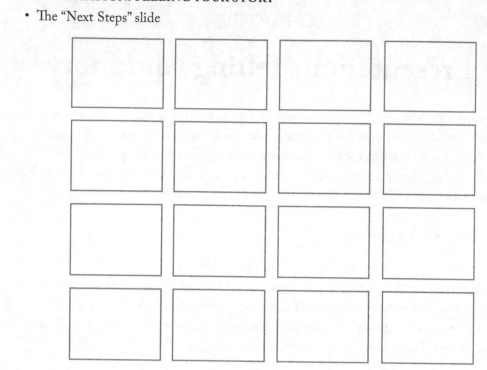

Figure 4.1: Storyboarding template to use to check your logical flow.

In different courses and different organizations, expectations of the types of slides vary. For example, faculty may want their students to omit a summary and next steps slide from a very brief presentation.

See Figure 4.2 for an actual example from an industrial engineering student group's proposal. The students have omitted the "summary" and "next steps" slide.

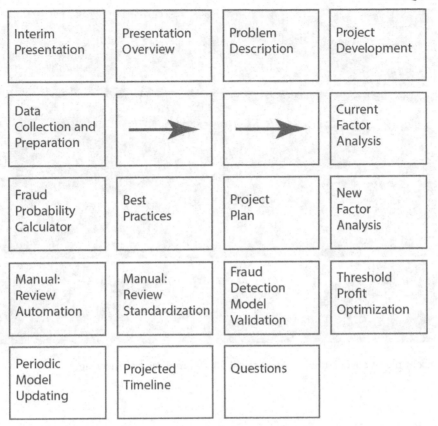

Figure 4.2: Example of storyboarding template used by capstone design group in industrial engineering.

The *cover slide* of a presentation should include a clear concise heading leaving no question about the topic to be discussed. For example, "The Capstone Design Interim Report" tells us very little, as opposed to "Procurement Planning for (an energy company)." The second title delineates the topic to come. Your cover slide should always include a date so if the slides circulate among different company employees they will be aware of the timeliness of the information. Often major participants in the project are listed. The company's logo and background may be used. An example of a cover slide appears in Figure 4.3 (from electrical and computer engineering), and Figure 4.4 (from mechanical engineering).

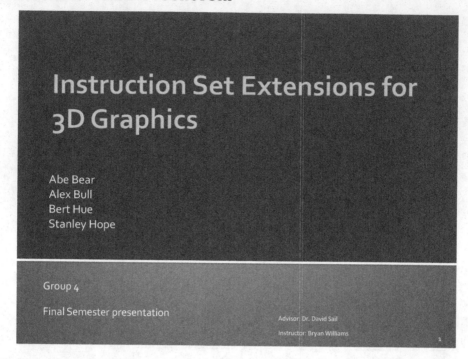

Figure 4.3: Example of cover slide from Electrical and Computer Engineering.

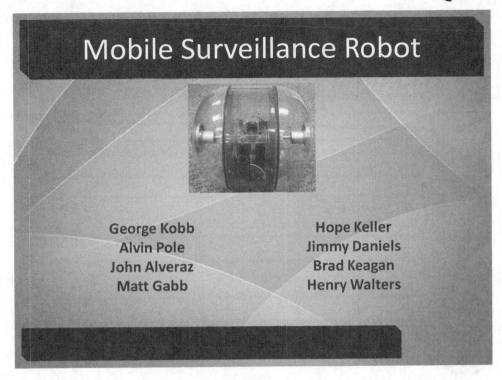

Figure 4.4: Example of cover slide from Mechanical Engineering.

Next, the *overview slide* explains, very briefly and often with the use of visuals:

1. the purpose of the talk and the related project, including the problem or "issue" being improved,

2. how the purpose was fulfilled, for example, what steps were taken in what order or what methods were used, and

3. the results of the project, often including a concrete description of the change (such as "an increase in efficiency of 30 percent") and the "value-added," or the money the project saved the organization.

Some beginning presenters wonder why they should describe their results right up front like this instead of including them at the end of their presentation. Many executives from a variety of settings stressed the old business adage that originated with George Bernard Shaw: "Tell them what you're going to tell them, tell them, and then tell them what you told them." The overview, or executive summary, engages the audience by "telling them what you're going to tell them"—letting the audience know exactly what's coming up. The executive summary is used by executives to determine if what they need to know is in the presentation. Examples of effective overview or executive summary slides are shown in Figures 4.5 (from Industrial Engineering) and Figure 4.6 (from Environmental Engineering).

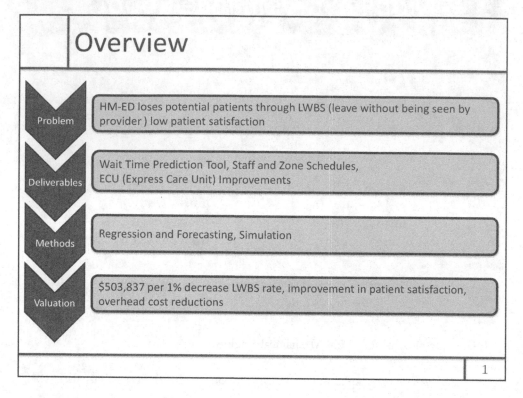

Figure 4.5: Sample introduction or executive summary slide from Industrial Engineering (ED stands for Emergency Department).

Figure 4.6: Example overview slide from Environmental Engineering.

After the executive summary, include *introductory* or *background slides* with a description of the organization related to the presentation (since many corporate names provide no idea of the company's business) and the background of the main problem with a description of relevant vocabulary. If your presentation is the second or third in a series of presentations, include necessary information from your past presentation.

The *main body* of your presentation will include the method used, the steps taken with an explanation of the need for each step, and the result or deliverables. For example, in some industrial engineering capstone design presentations, the main body consists of the data collected for analysis, the steps taken to analyze the data, and the deliverables and recommendations. Next include a *summary*, including no new information, to "tell them what you told them." Use a conclusion instead if you want to report the new information discovered in or resulting from the project. Remember, though, most of the executives suggested using the summary instead.

If your presentation is part of continued interaction with some audience members, end your presentation with a *Next Steps* slide. Presentations are almost never given in a vacuum—usually the presenter follows up with their audience to answer questions or resolve outstanding issues. Use the Next Steps slide as a starting point for discussion of future work to ensure you share the audience's expectations.

Check your slide titles to make sure you avoid using the same title for several slides. If you do, consider one of the following:

- Change the titles to represent the exact information on each slide.

- If you want to keep the same heading for each slide, distinguish each slide with a unique subheading. For example, "Method: xx Approach" and "Method: yy Approach."

- At the very least, use the phrase "(cont.)" on the second of two slides with the same title.

Exercise 4.1. Create your own storyboarding chart using boxes on a plain sheet of paper or a program. Enter each title from each draft slide into a box. Check whether your storyboard has the kinds of slides described below, in the order below.

Have you included an overview or executive summary slide?

Yes ___ or No ___

Have you included an introduction/background slide or set of slides?

Yes ___ or No ___

Have you included the main content? What chunks of information does it consist of?

Yes ___ or No ___

Have you included a summary slide at the end of your presentation?

Yes ___ or No ___

If your goal is to continue discussion (for example, with clients), have you included a "Next Steps" slide at the end?

Yes ___ or No ___

Finally, to check for the overall logical flow: have you provided enough information, before the audience sees the slides, to be sure the audience will understand the slides?

Yes ___ or No ___ (If you haven't, add the necessary information.)

For logical flow within each slide, check that everything on each slide is relevant to the slide title. Is it?

Yes ___ or No ___ (If it isn't, re-organize the information.)

4.1.2   MAKING TRANSITIONS

Executives told us that slide titles help build the story, and presenters should weave the story using transitions from slide to slide. Avoid simply reading the slide title and then discussing the main points on the next slide. Do not use repetitive transitions, such as "and now we'll talk about…" Instead, use transition sentences relating to the content.

For example, suppose one slide briefly describes all the steps in a process. If the next slide includes details on step 1, the presenter may say, "In step 1…." In another example, if slide 1 describes the output of a process which includes a model, the next slide might begin with you saying "the parts of the model include…" Remind the audience what you've covered so far and how this next topic fits into the big picture. You might say "now that I've described the results of the first step, I'll tell you how the results are used to create one of the deliverables: a program used to forecast the demand." Or, "Now that we've covered the types of data collected, we will describe the data analysis." The final phrase on one slide is or can be the lead-in to the next slide.

4.2   EMPHASIZING THE KEY POINTS

Consistently refer to how your key points fit into the big picture. Once I saw a presentation about baggage handling between planes of different airlines including exact pictures of the handwritten notes made by the baggage handlers. I can't remember what they did with the handwritten notes, how they analyzed them, and what proposed changes they made. The presenters forgot to remind us of how their details fit into their big picture.

Again, use transitions as opportunities to tie into the big picture. As one executive said, "to get or keep the audience's attention, refer back to what's in it for the audience." At regular intervals during the talk (at major junctions), remind your audience how the content you're describing fits into the big picture. For example, explain how producing a simulation representing cars moving through a car-auction site will lead to the project's goal. Include a small number of key points on each slide.

Exercise 4.2. List the transitions you used from slide to-slide in your last presentation.

1.

2.

3.

4.

5.

Can these transitions be improved? List several ways you can change your transitions that need improvement.

Idea 1.

Idea 2.

Idea 3.

4.3   CONTEXT

Clearly illustrate major points by linking them to additional relevant information, for example, stories or anecdotes. One executive emphasized using examples to put the solution in context. Use concrete examples to clearly and briefly explain every part of the presentation and to establish your credibility. For instance, show your audience how the outcomes of the presentation give them value—in their context. Support data with a concise verbal description. Explain the main point of the data and demonstrate, using a few details.

4.4   SENSITIVITY TO TIME

Stellar presenters respect their audience's time. Begin and end on time even with audience questions during the presentation. I was told of a presentation given by one engineer in a company to an audience of engineers. Despite the facilitator's indication that time was running out, the presenter said, "All right, I'll wrap this up quickly" and then continued to present slide after slide... I expect the audience saw the presenter as disrespectful of their time and they might have just stopped listening.

To end on time, take these steps.

Know the time limit before you present. Time your presentation when you practice, including time for questions.
Brainstorm possible questions before the presentation so you can give brief, accurate answers. Include appendix slides you may need to answer questions about details.
If during the presentation you're worried about finishing on time, first, decide quickly what can be left out and skip those slides, or ask the audience member with highest authority how they would like to proceed.
One executive summed up his approach to ending on time: "When you've made the sale, shut up!"

Several executives said that they often tell presenters to skip to the part most important for them. So you need to be prepared to let go of some slides and start on the requested segment. Focus on what your audience wants and needs to hear.

Exercise 4.3. Do you usually finish your, or your group's, presentations on time even when the audience interrupts with questions? Yes ___ No ___

If your answer is "no," then list actions you can take to be certain you will end on time.

1.

2.

3.

CHAPTER 5

Presentation: Displaying Key Information

Graphics and written information on the slides reinforce oral delivery through a focus on key points and supporting information. As the speaker, it is your job to make your slides easy for the audience to understand. Your listeners should not have to work to understand the slides. In this chapter, we review how to develop and design effective slides. First we show you three very effective slides, listed in the first table below. Then, in the second table below, we provide an overview of the rest of the chapter by listing the various types of slides we will discuss. All of the slides were created by Capstone Design students in various engineering disciplines in different U.S. universities.

Type of Effective Slide	Engineering Discipline
Slide with bullets and pictures	Biomedical Engineering
Slide with two graphs	Electrical Engineering
Slide with flow chart	Environmental Engineering
For a preview, the other slides included in this chapter are listed below	
5.1 Layout and Design	
5.1.1 Slides with Bullets and More	
Slide with bullet points	Mechanical Engineering
Slide with bullet points and diagrams	Mechanical Engineering
Slide with bullet points and map	Environmental Engineering
Slide with bullets and a flowchart	Industrial Engineering
Slide with bullets and a line graph	Industrial Engineering
5.1.2 Slides with Color	
Slide with chart including color on minimal amount of chart	Civil Engineering
Slide using light text on dark background	Mechanical Engineering
5.1.3 Slides with Information Highlighted	
Slide using call-out boxes (a PowerPoint element of graphical design) to highlight a key part of a stream	Environmental Engineering
Slide showing enlarged screenshot	Industrial Engineering

Slide including number chart with highlighting	Mechanical Engineering
5.2 Focused Content	
5.3 Amount of Text	
5.4 Engaging Graphics: Charts and Graphs	
5.4.1 Slides with Charts	
Slide with easy-to-read labels on bar chart	Electrical Engineering
Slide with flowchart	Chemical Engineering
Slide using flowchart for a timeline	Industrial Engineering
Slide including number chart	Electrical Engineering
Slide including pie chart	Environmental Engineering
5.4.2 Slides including Graphs and other Graphics	
Slide with diagrams	Mechanical Engineering
Slide with equations	Environmental Engineering
Slide with line graph	Industrial Engineering
Slide with map	Industrial Engineering
Slide with pictures	Mechanical Engineering
Slide with two graphics	Industrial Engineering

For each slide, the characteristics making the slide effective are noted. Now we return to the three effective slides.

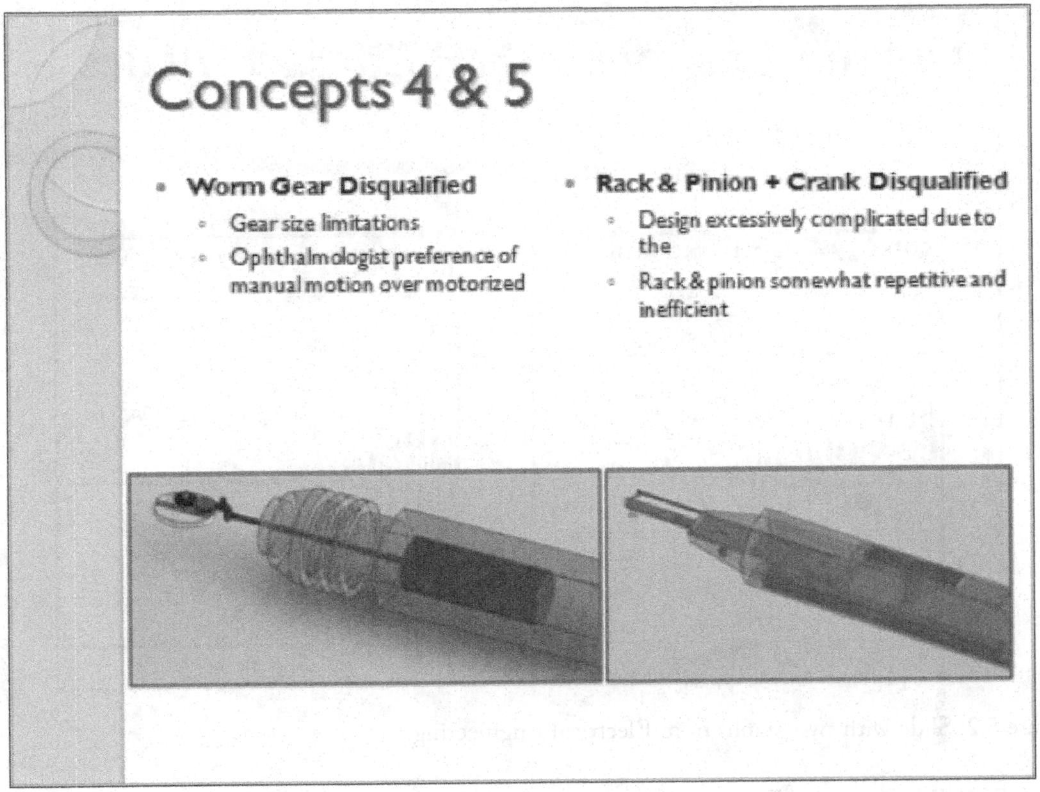

Figure 5.1: Slide with pictures from Biomedical Engineering.

Here are some of the reasons this slide is effective:

1. The title is clear and based on information presented before this slide.

2. Full sentences are avoided; phrases are used instead. If full sentences were used on most of the slides, the audience would read the sentences instead of listening to the speaker. And often the speaker would start reading the sentences on the slide, using less eye contact.

3. The pictures add value for audience members who have not seen concrete examples of the two concepts.

4. Each picture is labeled, with key characteristics provided.

5. The information presented in the slide is balanced.

6. The information is laid out so the audience can easily follow it.

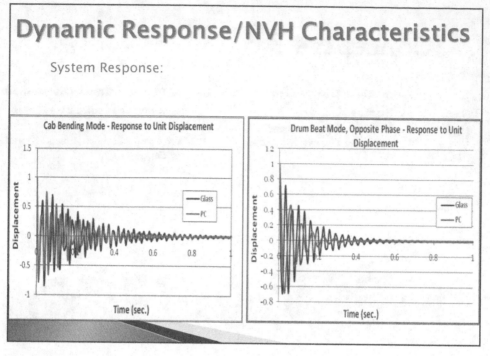

Figure 5.2: Slide with two graphs from Electrical Engineering.

The slide above is effective because it has the following characteristics:

1. The title is clear for electrical engineers. (Recall this slide is from Capstone Design for Electrical Engineers.)

2. The axes are clearly labeled on each graph.

3. The two graphs are clearly included on one slide for comparison purposes. If the graphs were included on two different slides, the audience would find it harder to compare them.

4. The keys for both graphs are easy to read.

5. The layout of the slide is balanced.

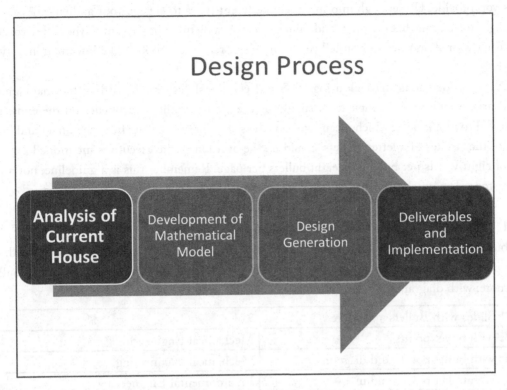

Figure 5.3: Slide with flow chart from Environmental Engineering.

Here are some reasons for this slide's effectiveness.

1. The title is clear.

2. Color and bold are used for emphasis. The last box is in color since it includes the results of the design process. The first box has bolded text, in color, because the speaker starts by referring to this box.

3. The labels are clear, concise, and easy to read.

4. The arrow underneath the boxes shows the order of the steps.

5. The text is large enough to be readable from the back of a large room.

5.1 LAYOUT AND DESIGN

When you use excellent layout and design, the information on your slide will be easily understood. The layout will be balanced, color will be used appropriately, and the key areas will be highlighted. Include on your slide the data needed to support your key points, and cues to additional explanations you provide. Use enough information of different types to engage your audience. Avoid including information that does not "add value"—such as a picture not relevant to the slide's content. Try using your slide titles to help tell your story. Also, remember to keep the information on your slide balanced.

Each slide should have clear logical flow. Keep the slides simple, without too many graphs and charts. Standard and simple fonts should be used: those without curly-cues on the ends. This is referred to as *sans serif*, which means without the short lines stemming from the upper and lower ends of the strokes of a letter. Bullets should not be overdone—the executives mentioned a guideline of eight words per bullet and eight bullets per page. Remember this is a guideline, not a rule set in stone. Too much information on slides obscures the main points for the audience.

5.1.1   SLIDES WITH BULLET POINTS AND MORE

Use bullet points to generate discussion of key ideas. The most effective set of bullets starts with the same part of speech, such as an adjective or verb. Below you will see examples of slides with bullets and more: with diagrams, a map, a flowchart, and a line graph.

5.1.1 Slides with Bullets and More	
Slide with bullet points	Mechanical Engineering
Slide with bullet points and diagrams	Mechanical Engineering
Slide with bullet points and map	Environmental Engineering
Slide with bullets and a flowchart	Industrial Engineering
Slide with bullets and a line graph	Industrial Engineering

In Figure 5.4, each bullet point starts with an adjective. The background makes the slide interesting but doesn't interfere with reading the words because of their large font. The amount of words on the slide makes it easy for the audience to take in, while the presenter has cues to speak from and then elaborate on these cues.

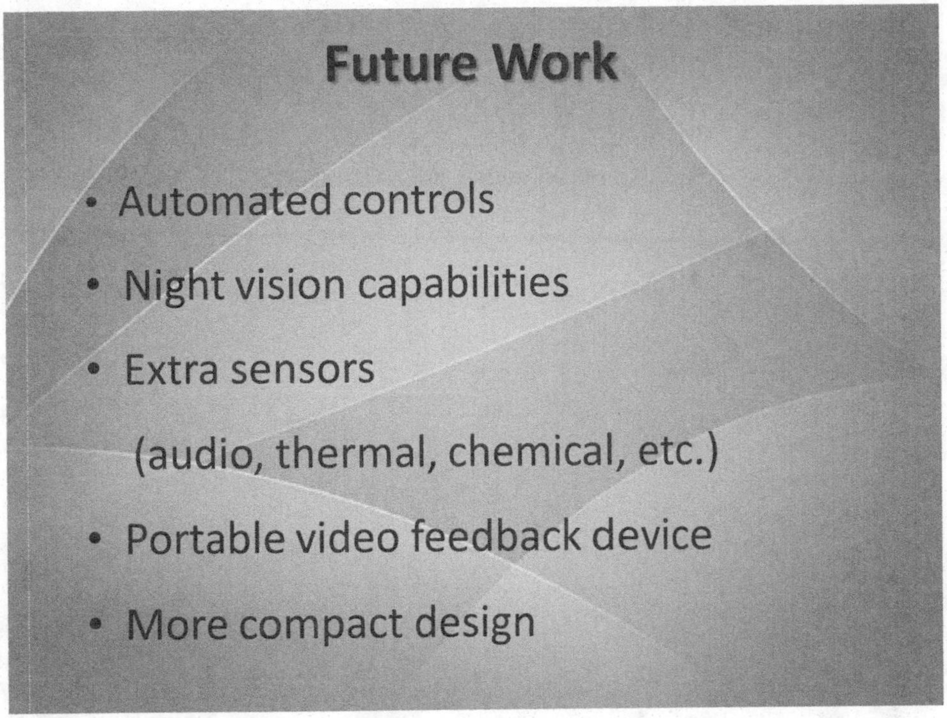

Figure 5.4: Slide with bullet points from Mechanical Engineering.

Next is an example of a slide using bullet points and diagrams in Figure 5.5.

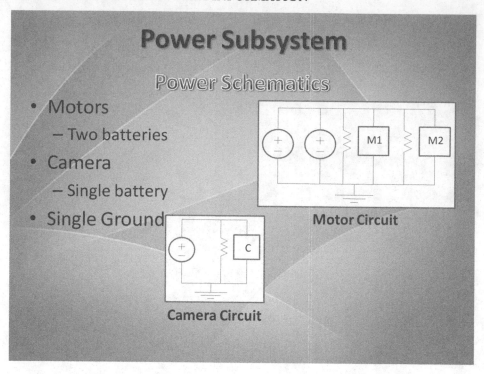

Figure 5.5: Slide with bullet points and diagrams from Mechanical Engineering.

Exercise 5.1. Describe four or five characteristics that make the above slide effective.

1.

2.

3.

4.

5.

On the following page you see an example of a slide including bullet points and a map in Figure 5.6, from Environmental Engineering.

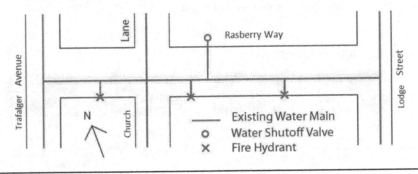

Figure 5.6: Slide with bullets and map from Environmental Engineering.

Exercise 5.2. The purpose of Figure 5.6, shown by its title, is describing the existing water system. Look closely at the map on the slide. Using this information, describe the water system by answering these questions.

1. Which streets are the end points of the existing water main?

2. Where is the water shutoff valve located?

3. How many fire hydrants are shown?___ Are any of them on Church Lane or Lodge Street?
   Yes ___ No ___

4. Was the slide effective in describing the water system? Yes ___ No___

The following Figure 5.7 includes bullets with a flowchart, and Figure 5.8 includes bullets and a line graph. Both slides are from Industrial Engineering.

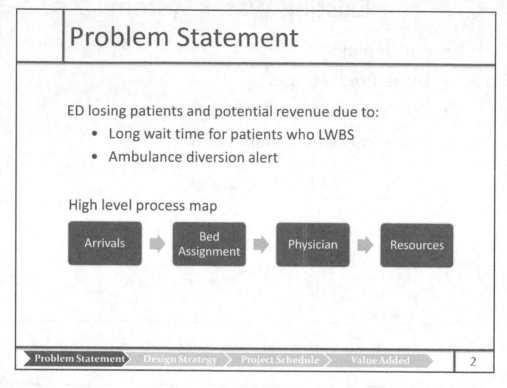

Figure 5.7: Slide with bullets and flowchart from Industrial Engineering (ED stands for Emergency Department, LWBS refers to a patient "leaving without being seen" by a provider).

As you can see, in this slide the concise bullets are used to explain the reasons for the main problem. The flowchart then shows the process used to assign patients to beds. The presenter noted that LWBS stands for "leave without being seen (by a doctor)," since this acronym would not necessarily be familiar to the audience of Industrial Engineering faculty and students.

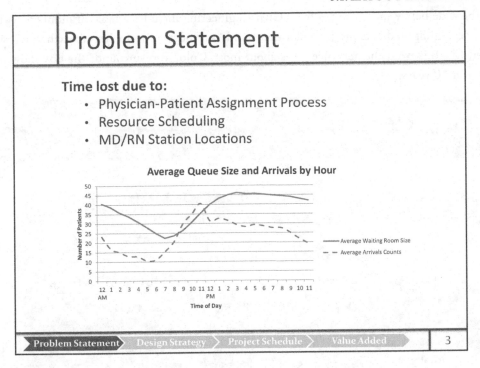

Figure 5.8: Slide using bullets and line graph from Industrial Engineering.

Figure 5.8 shows the use of bullets and a line graph. This slide actually appeared in a student group's presentation right after the slide in Figure 5.7. See how more detailed information is provided about the problem. The line graph is a visual that helps the audience see quickly the difference in two variables: average waiting room size and average arrival counts.

5.1.2   SLIDES WITH COLOR

Now that we've covered the layout of slides with bullets, we'll turn to more particular aspects of layout and design: the use of color in your slides, and highlighting. Color should be used sparingly, to emphasize critical text, numbers, or areas of graphics. Identify the slide's purpose and then select the key information. Highlight what you want the audience to take away from the slide. The slides included in this part are shown below.

5.1.2 Slides with Color	
Slide with chart, including color on minimal amount of chart	Civil Engineering
Slide using light text on dark background	Mechanical Engineering

The slide below in Figure 5.9 from Civil Engineering includes a table of numbers with different areas highlighted in color. The color that stands out most is the analysis of each year's savings and cost, which is what the audience cares about most. Color also surrounds the box showing the savings over 10 years.

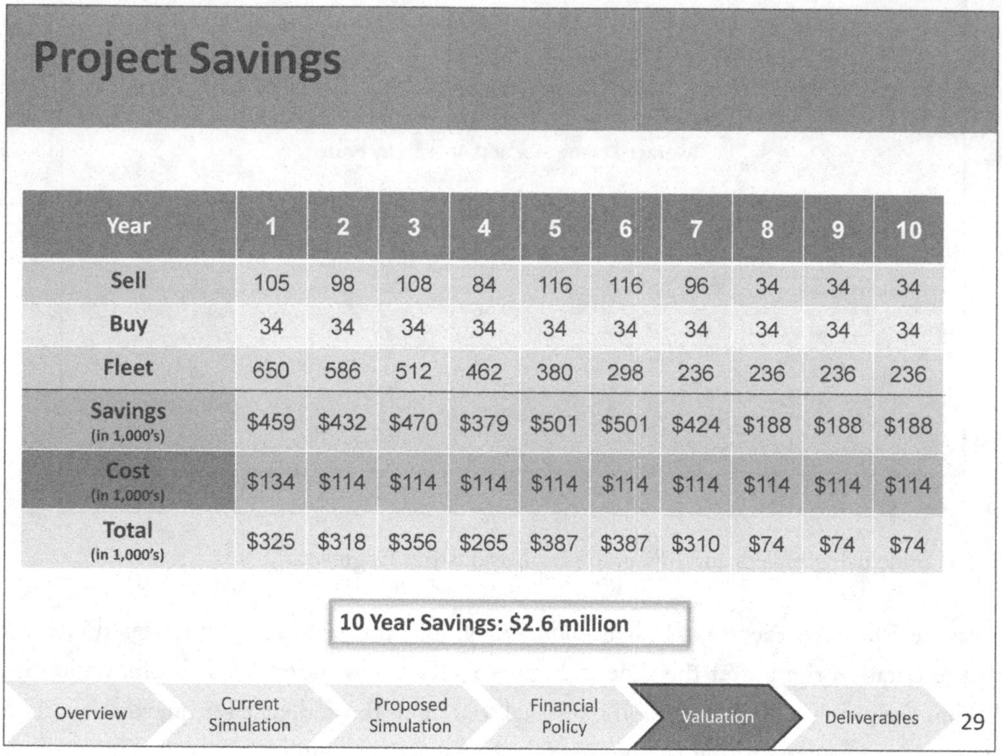

Project Savings

Year	1	2	3	4	5	6	7	8	9	10
Sell	105	98	108	84	116	116	96	34	34	34
Buy	34	34	34	34	34	34	34	34	34	34
Fleet	650	586	512	462	380	298	236	236	236	236
Savings (in 1,000's)	$459	$432	$470	$379	$501	$501	$424	$188	$188	$188
Cost (in 1,000's)	$134	$114	$114	$114	$114	$114	$114	$114	$114	$114
Total (in 1,000's)	$325	$318	$356	$265	$387	$387	$310	$74	$74	$74

10 Year Savings: $2.6 million

Overview ▸ Current Simulation ▸ Proposed Simulation ▸ Financial Policy ▸ **Valuation** ▸ Deliverables   29

Figure 5.9: Slide with table including color to highlight different parts of data from Civil Engineering.

The use of red: When selecting color for parts of your slides, use red sparingly because it is universally recognized as the color of emergency or crisis. However, if you use the color in small amounts, it definitely pulls the eye. For example, replace large blocks of red with boxes outlined in red. Avoid using red for slide headings unless you have a specific reason. For example, in one bottling company, red headings are part of the standard template because the company's logo is red. So if you are presenting to their officers, if they have asked you to use their template, you know they are comfortable with more red in the slides.

Also, be careful to use colors that "go together." Red and green, when side by side, can irritate the audience viewing the slide. I tell my students combining the two colors works for Christmas, but often not on slides. Slide designers want to show the negative in red and the positive in green. Consider trying another color like blue and bolding the text. To see if your colors complement each other, put the slides up using the same computer and the same projector in the same room where you will present. Each projector shows the slide colors a little differently; usually they are not the same colors as the ones on your computer screen. For example, yellow may become dark mustard—which can distract your audience.

Stay away from a large number of different colors so your viewer will not be distracted from the main message. One exception is when each part of the slide is important for a different purpose. To be effective in this situation, highlight the critical information on the segment you are talking about, while fading the other segments.

When selecting or creating a template, choose a background color that does not interfere with the message. Generally, light text on a darker background (as shown in Figure 5.10) is harder to read than a dark text on a lighter background. When you look at the slide below, you will feel the pull of the dark background, as compared to slides in the figures above. If all your slides have a similar dark background, your audience's eyes may tire over the length of the presentation. So, if everything else is the same, try using a medium or light color as background, and dark text.

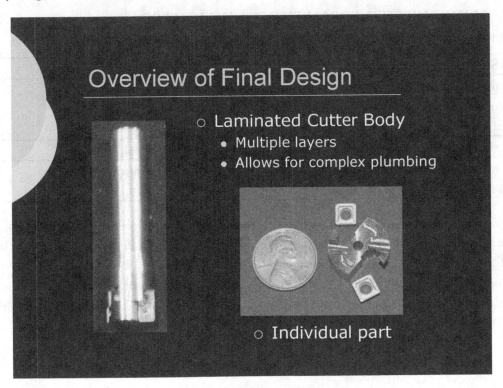

Figure 5.10: Slide with dark background from Mechanical Engineering.

Exercise 5.3. Design a slide using color, with any of the formats we have reviewed so far. Write the colors on your computer screen on the left in the table below. Now show the slides using the projector in the room where you will be presenting or have presented. Then write any difference in color on the right side of the table.

Colors on the Slide on your Laptop or Computer Screen	Colors You See When Using a Projector to Display your Slide on a Screen
1.	
2.	
3.	
4	

5.1.3   SLIDES WITH INFORMATION HIGHLIGHTED

The slides we include in this section will show how to highlight key information in your slides.

5.1.3 Slides with Information Highlighted	
Slide using call-out boxes (a PowerPoint element of graphical design) to highlight a key part of a stream	Environmental Engineering
Slide showing enlarged screenshot	Industrial Engineering
Slide including number chart with highlighting	Mechanical Engineering

You may use a call-out box, or an "element of graphical design" in PowerPoint, to focus attention on a small but critical piece of information. Using a call-out box draws attention to the key part of the information. For example, call-out boxes can highlight each segment of a country, one at a time. And one piece of the figure can be greatly expanded and moved out in front of or above the figure so the audience can see it very clearly. In Figure 5.11 (from Environmental Engineering) a call-out box enlarges a key part of a stream.

Figure 5.11: Slide with a call-out box from Environmental Engineering.

Figure 5.12 includes an enlarged screenshot, a second approach showing key information without using color. The screenshot has been expanded so the audience can read the labels.

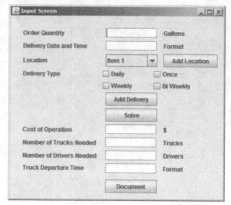

Figure 5.12: Slide with enlarged screenshot from Industrial Engineering.

If you use a picture of a screenshot so the audience can see exactly what your screen looks like, most likely the information will be too small for the audience to see or understand. Instead, use call-out boxes or enlarge a portion of the screen so your audience can read the information. Remember when audiences cannot see the words or information on the screen, they tend to ignore the speaker while they try to read the information, or they may tune out. Once this happens it is very hard to get their attention back.

The third example of highlighting appears in Figure 5.13. As you can see, key parts of the number chart are circled. These indicate the statistically significant factors—those with $p<.05$.

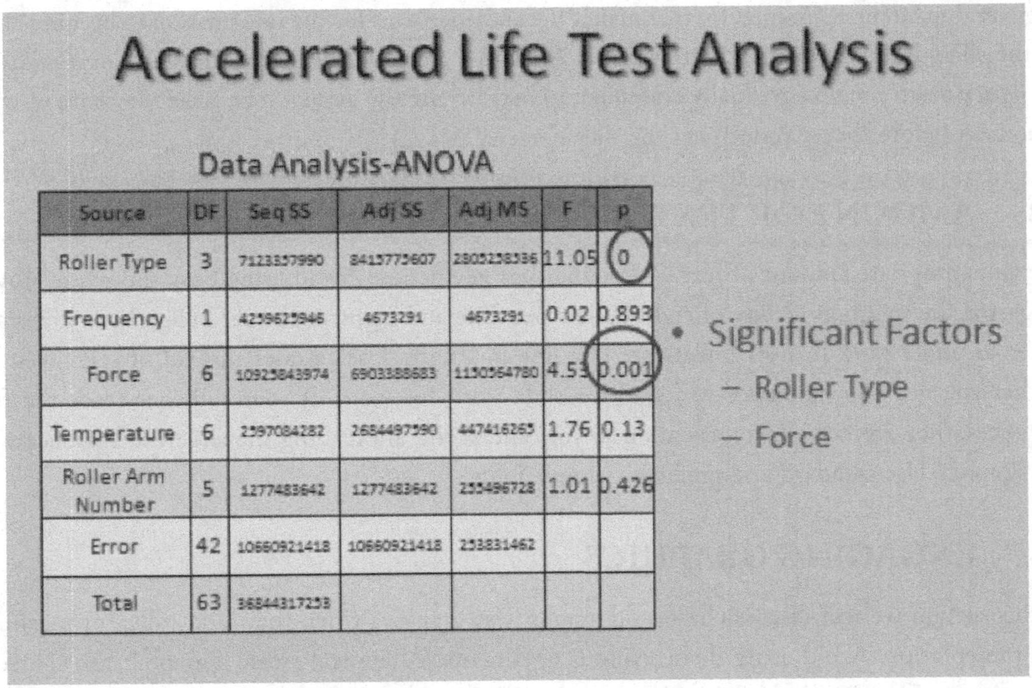

Figure 5.13: Slide with highlighting on number chart, from Mechanical Engineering.

5.2   FOCUSED CONTENT

The information on a slide should support only one or two main points—keep the content focused. Supporting or illustrating information may be included as well. To check your slides, take on the role of your audience, and ask yourself, "If I'm seeing the slide for the first time, will I be comfortable with all this information or will it make me want to withdraw?"

Use minimal animation. Learn to be selective to avoid distracting your audience.

One slide presentation I saw had a yellow truck on the lower left side of the presenters' cover slide. (Their client was a delivery company.) As they presented, the yellow truck slowly moved from left to right, stopping on the bottom right side of the slide. I was trying to listen, but I could not keep my eyes off the colorful truck until it stopped. By that time I had missed most of the presenter's discussion.

Instead of this approach, you might include the moving truck on the final slide for Questions, when it would not interfere with the audience's understanding. Or you could use the truck only on the cover slide, and keep it static.

Chunk information carefully. Bringing text in one chunk at a time helps viewers process the information. This approach limits the amount of information the audience needs to process at once.

In general, avoid using animation that makes the audience wait for the information to become clear on the slide. *Fly in animation* (with the text "flying in" from both sides of the slide) or *dissolve in animation* (with the text gradually crystallizing) may irritate the audience because they have to wait a moment before they can decipher the slide.

5.3   AMOUNT OF TEXT

Use an appropriate amount of text to describe your key points. Avoid using busy slides; executives suggested the rough guideline of eight bullets per slide and eight words per bullet. Your slides are easier to understand if most bullets are one line in length. Use phrases instead of sentences, so neither you nor your audience will read the slide word-for-word. If your audience reads the full sentences, they are not paying attention to you, and if you are reading the slide the audience will feel ignored. Use standard and preferably simple fonts.

5.4   ENGAGING GRAPHICS

In this section we will discuss charts and graphs. Use graphics when they "add value" or meaning to a presentation. Avoid using them solely as ornaments. When you create a graphic, ask yourself, "what is the main purpose of this chart or graph?" Use *only enough information or data to get your point across*. Often engineering or science students and professionals want to show all the data they collected to reveal how much work they've done. But they may end up with an audience whose eyes are glazed over.

Engaging graphics need readable labels. Again, if an audience member has tuned out because they can't see the labels, you will need to work very hard to re-engage them. Some viewers will ask questions for clarification but some will not. Labels on graphics need to be in units the audience is familiar with—for example, U.S. standard measures should be used for audiences in the U.S.

Choose a type of chart that clarifies your main points. For example, a graph may show an increase in demand more clearly than a table. And a histogram or bar chart may indicate more clearly than a pie chart the difference in size of sales for products.

5.4.1 CHARTS

In this section, the following slides will be included:

5.4.1 Slides with Charts	
Slide with easy-to-read labels on bar chart	Electrical Engineering
Slide with flowchart	Chemical Engineering
Slide using flowchart for a timeline	Industrial Engineering
Slide including number chart	Electrical Engineering
Slide including pie chart	Environmental Engineering

Bar Charts. Figure 5.14 below shows easy-to-read labels on a bar chart used in Electrical and Environmental Engineering.

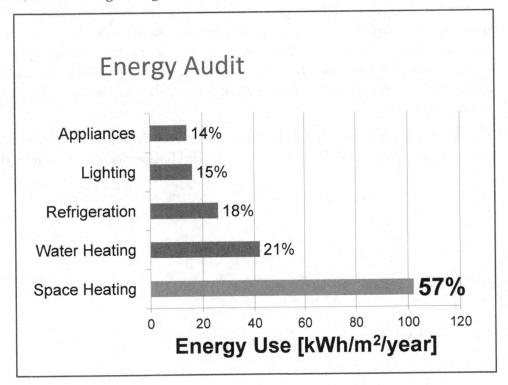

Figure 5.14: Slide with Easy-to-Read Labels on Bar Chart from Electrical Engineering

If you pull a chart from another piece of software (such as Visio—a professional diagramming tool, or Arena—simulation software) into PowerPoint, the labels are tiny. Re-create the graph in PowerPoint with bigger labels. Or highlight key pieces of the whole and then expand and view only one section at a time.

Flowcharts. Below are principles guiding design for flowcharts:

1. Keep the flowchart simple using only enough information to demonstrate the point.

2. The flowchart should support your description so you will need fewer words.

3. The words must be large enough for the audience to see.

4. The abbreviations used must be part of the audience's vocabulary.

The audience will follow you better if you highlight the boxes in the flowchart as you discuss them. Use a circle, bolding, color, or a callout box.

Some students who brought in slides with flowcharts from Visio told us they didn't care if the audience could read each label. They just wanted the audience to get a feel for how large and complex their work flow was. But, regardless of what the speaker wants, the audience will squint and try to read the labels, even though they are impossible to see. Better to use several large labels showing key points in the flowchart. Second, in some cases I've seen students present a non-readable flowchart and say, "I know you can't read this…" This behavior is actually an insult to the audience, who would tend to think, "Then why are you showing it to us and wasting our time??"

The next two examples use effective flowcharts. Figure 5.15 shows existing operations. The slide is from Chemical Engineering. Figure 5.16, from Industrial Engineering, shows a clear timeline.

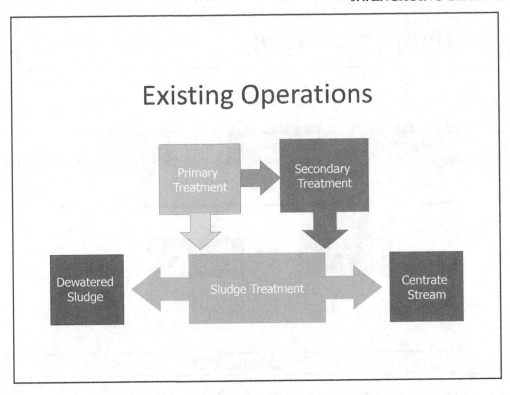

Figure 5.15: Slide with flowchart from Chemical Engineering.

Figure 5.15's flowchart is easy to read, with clear arrows showing flow, and clear labels. The same is true for Figure 5.16 below, where a flowchart is used to make a timeline.

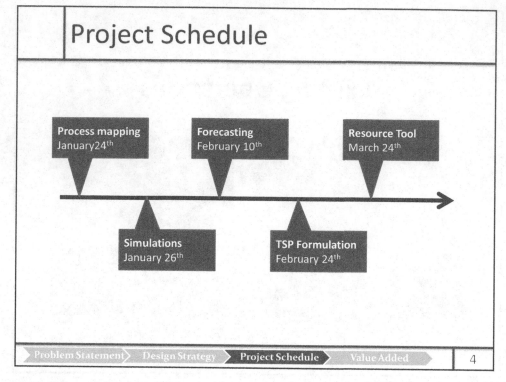

Figure 5.16: Slide with use of flowchart for timeline, from Industrial Engineering.

Exercise 5.4. For your own current project or one in the past, use the flowchart approach shown above to build a clear timeline. Use the box below to design the slide.

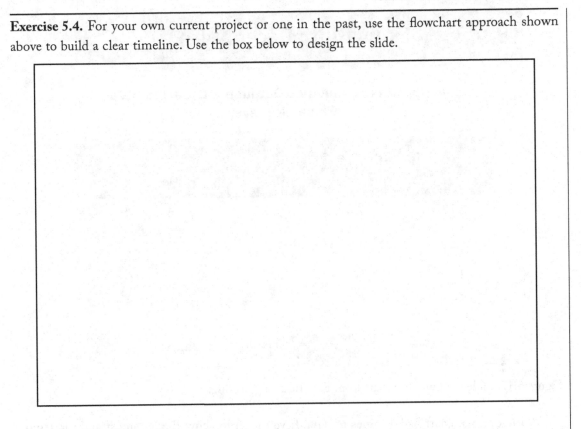

Number Charts. Design number charts to display only key information, but enough to help the audience make sense of the resulting numbers. Use the number of significant digits expected by your audience. The slide from Electrical Engineering in Figure 5.17 includes a chart with enough numbers to get the point across. The audience will find this number chart easy to understand.

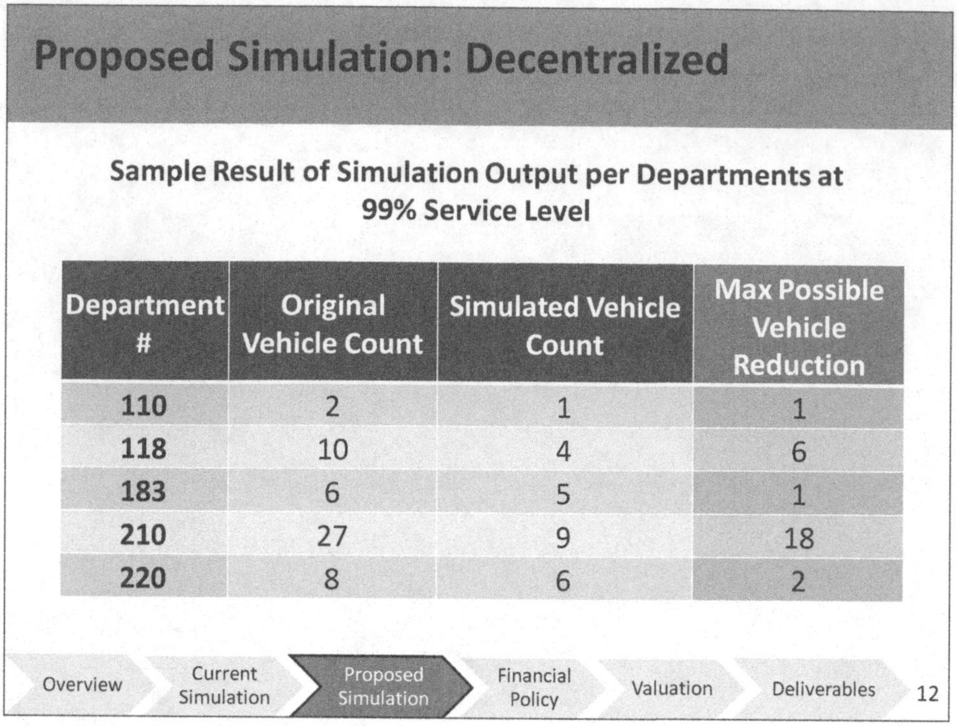

Figure 5.17: Slide with number chart from Electrical Engineering.

The original chart had 12 lines of data. If you need to show the left-most and right-most columns, use "…" in place of center columns. Sometimes complex flowcharts or tables are necessary. In one case in Industrial Engineering, students needed to show the results for all 15 of the company's products. The chart dealt with the products individually. If the information had been split into two slides, the impact would have been reduced. In this case, highlight a piece of the table and expand the section in chunks that "pop in" instead of "fly in" or "crystallize." You will explain best if you control the speed of the animation you use.

Pie charts. A pie chart is shown in Figure 5.18 (from Environmental Engineering). They are useful for showing the relative size of pieces of a whole. In this figure, many materials are listed—necessarily, because each contributes to the whole. Different colors in the pie chart match the list of materials in the chart on the left.

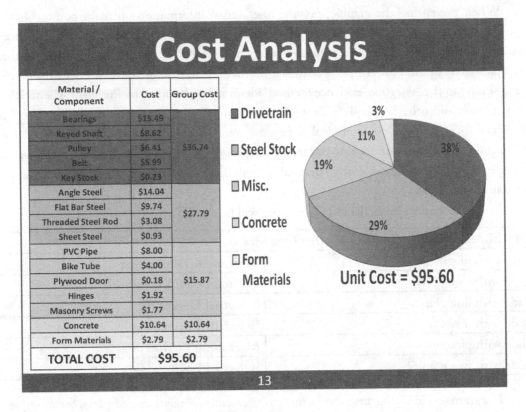

Figure 5.18: Slide with pie chart from Environmental Engineering.

5.4.2 SLIDES WITH GRAPHS AND OTHER GRAPHICS

Use visually appealing and easy-to-understand graphics, such as diagrams, line graphs, and maps, with helpful labeling. An engaging graphic is "easy on the eye," or pleasant to look at. For example, the colors used to illustrate main points should not clash. The labels associated with the graphic should be easy to see and in a language audience members can understand. Some engineering students work with systems that have "go" or "no go" conditions. In this case, instead of using red and green, consider red and a complimentary color, or black and white.

When creating graphics, first clarify the purpose of the graphic and then include only enough information to get the point across. All of your information should support the purpose of the graphic. Your audience must be able to recognize the most critical part of the graphic. Focus much of your preparation time on your graphics. They are often used to describe the results of a project, which determine whether or not the project was successful and what the next steps should be. Pare down the data to allow the main points to shine. You will see multiple examples of engaging graphics below, starting with diagrams and continuing with equations, line graphs, maps, and pictures.

When presenting the graphic, provide brief verbal information on main points. Once or twice I've seen engineering students present graphs or charts in silence, expecting the audience to identify the main point. This makes the audience work, instead of you doing the work for them. Some presenters opt to have the purpose in text on the slide next to or above or below the graphic. This is similar to the "assertion-evidence method" identified and studied by Professor Michael Alley at Penn State and others. This slide design can be very effective, especially if you are describing technical information to a non-technical audience. Even if the purpose of the graphic is spelled out on the slide, describe it and include concrete examples of the main point. As one executive said, "The speaker should summarize why I'm looking at all these numbers. They must summarize using one main point."

In this section slides of the following types will be shown.

5.4.2 Slides including Graphs and other Graphics	
Slide with diagrams	Mechanical Engineering
Slide with equations	Environmental Engineering
Slide with line graph	Industrial Engineering
Slide with map	Industrial Engineering
Slide with pictures	Mechanical Engineering
Slide with two graphics	Industrial Engineering

Diagrams. A drawing that shows the arrangement and relations (of parts, for example) is a diagram. The purposes of diagrams vary. They are used frequently in many types of engineering. Figure 5.19 displays the parts of the final design of a W-axis assembly. This was created by Mechanical Engineering students.

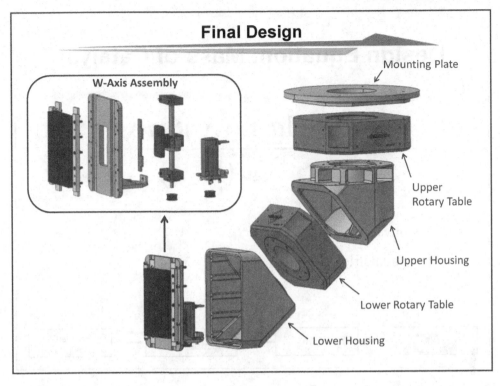

Figure 5.19: Slide with Diagram from Mechanical Engineering.

Equations. Engineering and science students and professionals use equations. If the variables are not familiar to your audience, describe each one on your slide. An example is shown below in Figure 5.20, from Environmental Engineering. Note that the variable definitions are easy to read.

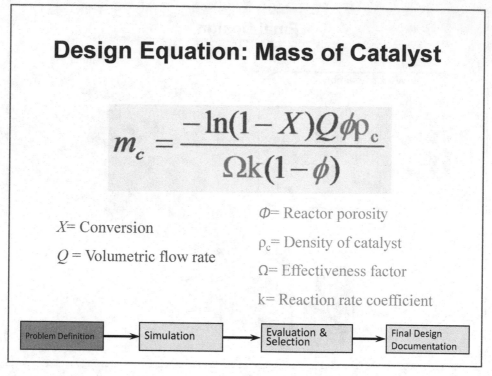

Figure 5.20: Slide with Equation from Environmental Engineering.

Line Graphs. Use line graphs to show information in a straightforward and easily digestible way. For example, if you have seven lines on your graph, your audience will be overwhelmed and stop paying attention. Consider the relationships you want to depict, and use the relevant lines to show these. Revealing the lines on your graph one by one will give you time to explain what each line represents and how it contrasts with the others. This approach also gives your audience time to ask questions about the data. If you have three lines on the graph, in different colors, and a key that is easy to read, you will not need animation.

Use line graphs to display the differences between two different sets of data. In Figure 5.21 (Industrial Engineering), cost per year is contrasted to salvage value per year.

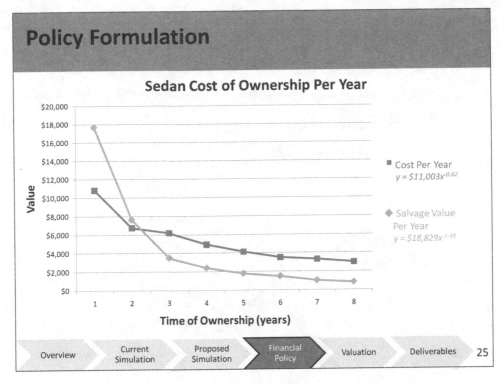

Figure 5.21: Slide with Line Graph from Industrial Engineering.

Maps. When you use maps in slides, be sure they are easy to read and clearly labeled. Identify why the map is included. In Figure 5.22, from Industrial Engineering, individual areas in the map are circled and described with a box with large labels. The map is easy for the audience to see and understand.

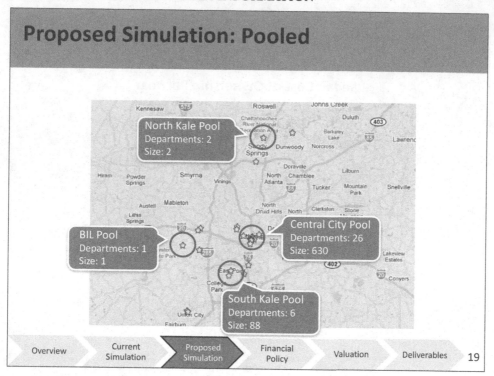

Figure 5.22: Slide with map from Industrial Engineering.

Include a **picture** if it provides new information for your audience. In a capstone design group with a real-world client, often only a few members of the academic audience have visited the client. In this case, a picture of the company's site and their products will provide important information. Do not use a picture that will distract the audience from the main message. For example, years ago one capstone design group worked with a large group of dental professionals. On each slide in the heading portion, they included a close-up of a patient's head, while they were sitting in the dental chair, with the dental professional leaning over their very swollen red gums with a dental tool. I cringed when I saw the picture. In contrast, pictures showing the inside of a warehouse are informative. When adding a picture to a slide, use a size that does not distract or overwhelm the audience.

Use pictures to add information best understood through visuals. Whether or not the pictures are labeled, add a verbal description as you present each slide. Figure 5.23 shows pictures used during one Mechanical Engineering project.

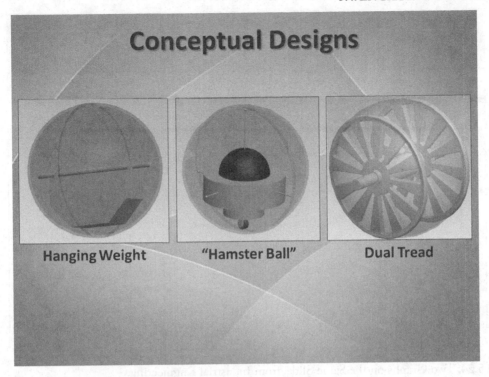

Figure 5.23: Slide with pictures from Mechanical Engineering.

Slides with Two Graphics. Sometimes two graphics on one slide, with easily understood labels and units, help the audience compare or contrast similar information. However, if more than two graphs are used on one slide, the audience may not see any graph clearly. The audience has real trouble understanding slides that include three or four graphics. In Figure 5.24 you see an example of a slide with two graphics, clearly labeled, from Industrial Engineering.

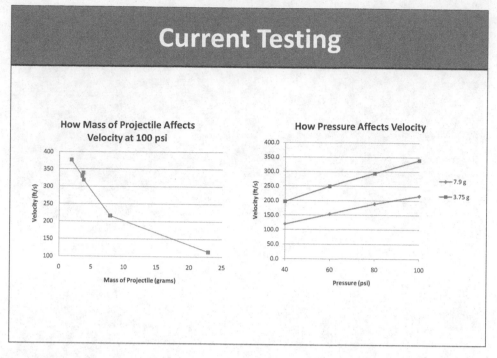

Figure 5.24: Two Graphs on the Same Slide, from Industrial Engineering.

Exercise 5.5. Describe the main message of this slide and the slide's characteristics that support and clarify this message.

Main message:

Characteristics supporting main message:

1.

2.

3.

4.

Now that we have covered how to create and display key information, we turn to suggestions for delivering the presentation.

CHAPTER 6

Delivering the Presentation

Keep in mind that some things may be out of your control during your delivery. We have all seen a presenter whose slides stopped working or whose computer failed during the presentation. Sometimes the fix is not easy or quick. Do some "disaster planning" so you will have a plan whenever you need to present without slides.

Connecting with the Internet may be a problem. So carry with you a flash drive or CD with a backup copy of your slides.
Your audience will take its cues from you on how to respond to the situation. If you stay calm and continue with your backup plan, you will keep their attention.
Ahead of time, be clear about the main points you want your audience to take away from your presentation. When the glitch happens, if you cannot resolve the technical issue, continue your talk by covering the main points. The details will come up in later discussion.

One principle guideline for delivering presentations cannot be stressed enough! It is your job as speaker to do the extra work to make sure your message is easy for the audience to understand. Avoid making the audience work to understand. This chapter includes six specific sets of verbal and nonverbal skills used by presenters to enhance the delivery of their message. The skills are: first/last impressions, flow, elaboration (or saying more than what's on the slide), stature (or posture), vocal quality, and personal presence.

6.1 FIRST/LAST IMPRESSIONS

Capture your audience's attention at the beginning of your talk and inspire them with the closing. This keeps your audience engaged.

Choosing the Content of the Presentation's Introduction and Closing. For your introduction, explain to the audience which of their needs the presentation will satisfy, and why the audience should listen. Often the audience members care most about the bottom line: what are the savings if we follow your recommendations?

Use the introduction to tell what the presentation is about in a very concise fashion. The overview basically answers "Who? What? Where? When?" and "Why?" Each time you present the project, describe:

• the problem or issue you faced at the very beginning of the project,

- the current situation, or the reasons you identified that caused or were related to the problem,

- the approaches you applied to fix the problem,

- the results of the approaches and the deliverables you provided to the client or person who came to you with the problem,

- the explanation of why the results resolved the problem, and

- future steps to take to continue your work, or when your project is done, to help your client implement your recommendations.

At the end of the presentation, inspire the audience to action—to follow up on your recommendations, for example. Use concrete information relevant to the purpose of your project. Although much of the information you cover in your introduction and summary will be similar, be sure to use different visuals on the slide to make your point. Summarize the work by "telling the audience what you told them" to help your audience remember the main points of your talk. End with a call to action or an update, or by providing the recommendations once more.

Delivering the Introduction and Summary. If you don't feel excited, act excited. I tell my students that, even though they have heard or given this presentation numerous times, act like it is the most interesting topic in the world. And I raise my voice and use inflection to get my point across to them. Here are a variety of ways to act excited:

1. Talk more loudly, or talk using high and low volume to emphasize key points and why the audience should care about your topic.

2. Talk more quickly, but not so fast the audience can't keep up with you.

3. Use hand movements that are more emphatic, such as opening both your hands slightly away from your body. This gesture brings the audience in and makes them feel included.

4. Make eye contact for longer than usual with many of the people in your audience.

5. Raise your eyebrows and open your eyes a little wider than usual.

Exercise 6.1. Ask a friend to listen to you, or look in your mirror, to practice the actions described above.

1.   Which of the five suggestions above are you using?

2. How can you improve your use of these suggestions?

3. Which one of the other suggestions would be helpful to implement? Note: try adding one when you are comfortable with the suggestions you are already using. Which suggestion would be helpful to capturing your audience?

4. How would you implement this suggestion?

5. Practice these actions, using different words to avoid memorization, until they become second nature to you as you open or close a presentation.

6.2   FLOW

Your flow is smooth if you know the material well but haven't memorized it, and if you avoid many hesitations or overuse of the word "um." These two situations will be described below.

6.2.1   AVOIDING MEMORIZATION

If you memorize your material, when you are interrupted, you'll forget where you were and have difficulty picking up where you left off. You may end up making several comments you've already made (as in the example below). When your audience sees you this bewildered, they may stop listening. They may also become disappointed or irritated. They came to the talk expecting a discussion, not to hear a memorized recital of facts.

> *During one Capstone Design student's final presentation a number of years ago, the student stopped in the middle of their sentence. I'm not sure what interrupted them or "set them off." I do remember the speaker was looking at the wall for a moment. After a very long interval, very noticeable by the audience, the speaker started talking again with the last two sentences they had already said, and then pressed on. Clearly the presenter had memorized a script, and once*

they were thrown off, they couldn't find their place and had to back up a ways to get re-oriented to their script.

Memorization is often given away by the speaker's lack of inflection or eye contact, or use of too many "um's" (as described below in 6.2.2). If a presenter says an acceptable phrase and then stops to "correct" themselves by saying a slightly different phrase with the same meaning, this usually means they have memorized. To avoid memorizing your talk, each time you practice, say your presentation slightly differently. Remember key points—with the help of your slides.

The following guidelines will help you avoid memorization or sounding memorized.

Never write any part of the talk down.
Never use notes.
Use your slides as cues.
Avoid over-practice. For example, one student group told me they had practiced their presentation 15 times. That's too much.
When practicing, use different wording each time.
Use inflection to reflect the most important points.

6.2.2   AVOIDING THE USE OF "UM'S"

Presenters often show hesitation or say many "ums" because they do not know the material. Be sure to be familiar with your material. Pausing every so often at natural points is fine, but let the pause be silent.

In our workplace presentation instruction, many times students will use a great many "um's" especially when practicing their presentation for the first time. When we work with a team, often the team members will notice the overuse of "um." In several cases, I've seen a team member actually count the number of "um's" the speaker said while they presented. But sometimes even this feedback isn't enough to convince the speaker that they are using so many "um's" that the audience will be distracted from their main message.

When a speaker isn't convinced that they use many "um's," a videotape of their presentation will convince them.

Usually when I ask the speaker if they can remember what they were thinking when they said "um," they say they are thinking about what they will say next. I suggest leaving a blank pause instead of using the "um"—this will also let the speaker think about what they will say next. I demonstrate a significant pause to the team, and they look at me waiting for what I will say next.

The speaker then selects one of their slides to use for practice. They start to present, and whenever they say "um" I stop them. If it happens on the second or third sentence, I ask them to back up to their first sentence. Usually, speakers will be able to avoid "ums" with a short enough piece of a presentation. Once they are successful, their teammates and I give positive feedback, and then ask the speaker to "go for it!" for the first two sentences. Slowly we build up the sentences that are without "ums," and then I ask the speaker's teammates to continue practicing with the speaker outside our Workplace Communication Lab.

Exercise 6.2. Practice to avoid too many "um's".

Use your phone to videotape yourself, or use your computer to videotape. If you have a friend who can join you, this will be even easier. Stand up with a slide where you need to add a lot more information than what appears on the slide. The slide can be on a computer, a poster, or a piece of paper.

1. Videotape yourself explaining the entire slide. Do you hear "um's" when you play the videotape back? Yes ___ No ___

2. If you do hear "um's" then back up to practice only the first sentence, using short pauses as described above. What do you notice about your "um's"?

3. If you hear no "um's" continue your practice with the one sentence until you feel comfortable. Then present the first two sentences, etc., adding more sentences when you have succeeded in avoiding "um's" so far. What do you notice at this point?

4. Describe the process that is working for you. For example, are you giving yourself lots of practice time once you reach three sentences?

5. Practice this regularly until you avoid most "um's." Be sure to change your words each time to avoid memorizing.

6.3   ELABORATION

Avoid reading every single word off your slide since you will insult your audience by ignoring them. If you don't elaborate, your audience will think: 1) you don't know your topic, or 2) you're embarrassed to look at them.

Instead, elaborate or add information not shown on your slides, for example, by describing real-life examples. Practice using the slides so you can glance briefly at the slide and then look at your audience. Stand next to the slide with some movement and avoid having your back to your audience. When the screen is very large, help the audience follow along by showing them, here and there, where you are on the slide. Refer to the main points on the slide, or some of the main points. Using a laser pointer is an option, or you can use phrases such as, "As you can see from the top point on this slide…."

6.4   STATURE

Your stature includes your posture and bearing. Bearing refers to how you hold yourself. Stand up and present confidently. The audience sees bad posture as unprofessional; leaning on the podium or slouching shows you're not serious. In contrast, good posture helps draw your audience in.

If you hold yourself and move as if you have something really important to say, your audience will respect you and see you as more credible and trustworthy. For instance, avoid picking lint off your outfit as you present.

6.5   VOCAL QUALITY

When you present with excellent vocal quality, you adapt your tone, volume, and pace to emphasize key points. Each of these is discussed below.

6.5.1   TONE

Use a tone that reflects confidence and appropriate intensity. If possible, use a lower pitch, since presenters with very high voices as perceived as less serious.

6.5.2   VOLUME

Speak loudly enough to be heard. The larger your presentation area, the more you need to project. One useful exercise is for a coach to say, in a very definite tone and manner, "I can be definite." Then a soft-spoken teammate is asked to repeat the sentence. Teammates provide feedback until by the second or third try the presenter does sound definite. Then the presenter can continue practicing with the help of their team.

Exercise 6.3. Check your volume before presenting in very large rooms or auditoriums.

Before you present in a very large space, find a similar space and place your phone on video-tape at the other end of the room—that is, at the very back. A few audience members will probably be sitting in the back.

1. Go to the front of the room and practice presenting, with your slides on your laptop. In this particular practice it is not important whether the video captures you; you will be checking to see if you can hear yourself clearly.

2. How did you do the first time? Were you loud enough? Yes ___ No ___. If not, continue practicing until you can easily hear yourself on the video. Expect yourself to sound extremely loud when you are projecting well, since you don't normally speak at this volume.

Use substantially higher volume as well as a pregnant pause to selectively emphasize key points. For example, when presenting a summary at the end of a project, speak louder when you discuss the value your project added. One of our Capstone Design projects focused on making a railroad more efficient and as a result saved the railroad $400,000 per year; other projects saved their clients up to $5 million annually.

6.5.3 INFLECTION

Avoid using a monotone; instead emphasize different words or phrases when you practice your talk. Use inflection to answer audience questions to show you are open to their input.

Be sure to go down at the end of each sentence, so your comments don't sound like questions. Be careful to keep your sentences reasonably short, with pauses in between, so the audience can follow easily.

6.5.4 PACE

Enunciate well, or speak distinctly when you present. Make sure the audience doesn't need to work to understand you. Check your pace by practicing in front of friends or colleagues in a situation as similar as possible to your actual presentation. Talk slowly enough to be understood.

A presenter with superb vocal quality uses a high enough volume, modifies their volume to emphasize certain points, uses a smooth and reasonable pace, and uses pauses appropriately.

6.6   PERSONAL PRESENCE

Personal presence represents the effective combination of energy, eye contact, and movement. A change in one often brings about a change in the others.

6.6.1   ENERGY

Stellar speakers get excited about the main points of their presentation. Show an avid interest in your message even though, because of practice, you are familiar with it. As the presenter, you are an expert on your topic: be strong and enthusiastic in your knowledge.

6.6.2   EYE CONTACT

This is a basic but all-encompassing part of presenting. If you are an "excellent presenter" except for your eye contact, you are really an average or poor presenter. The audience will not look at you if you don't look at them. So instead of looking up, or down, or at the ceiling or the wall, practice making direct eye contact with several teammates, ahead of time. If you are very shy and have trouble looking directly in the audience's eyes, 1) look at the audience's foreheads, 2) look between their eyebrows, or 3) focus on two or three audience members in different parts of the audience instead of surveying the whole room. Practice until you are able to make eye contact. Be sure to speak to your audience instead of to your slides. Do your best to look at individual members of your audience.

Use eye contact to check your audience for nonverbal cues indicating misunderstanding, negative reactions, or boredom. How you handle the response depends on the context. For example, when an audience member shows possible misunderstanding by a tilt of the head, a squint or a slight frown, ask if there's something needing clarification. If you see a negative reaction by some of the audience, shown by severe frowns or a jutting of the chin, discuss the issues in more depth by either taking time out from your talk or setting a separate time for discussion.

If you see an audience member not paying attention, consider asking for questions at the end of the segment, looking at but not singling out the bored audience member. Your response depends on the exact situation, so be careful to take in all the verbal and non-verbal nuances. When you address an audience of individuals with differing authority, spread your eye contact around instead of focusing mostly on the decision maker so others do not feel left out.

Avoid using notes. Your audience will perceive you as ill-prepared and disrespectful of their time. In their eyes, if you don't care enough to prepare well, why should they pay attention? Get to know your information well enough to use your slides as cues, and elaborate on them as you look at your audience.

6.6.3   MOVEMENT

Hand movement should be used for emphasis. Use your hands selectively, naturally, and confidently. Think ahead of time about the topics you want to emphasize.

Using your hands too much can distract your audience. Focus on reducing your hand movements instead of stopping them, since you may end up looking very stiff. Practice in front of friends or colleagues. Certain hand gestures are very helpful:

- "counting" on your fingers to enumerate points,

- spreading your arms to indicate a larger size, or

- using your hands to show three levels.

    Hand gestures that are distracting include:

- movements you repeat over and over without relating to your content,

- flipping your hair back repeatedly, and

- playing with the change in your pocket (the executives said this one was particularly annoying.)

Body movement—such as moving closer to your audience and then returning to your slide—is sometimes used by executives. Minimize the time the audience sees your back by:

- moving your body sideways, and

- taking a step or two up, and then back.

Some people see nothing wrong with walking across your slides or standing in front of your slides. Avoid standing in front of the slides so your audience is not distracted from your message.

Avoid swaying from foot to foot, since the movement is hard for your audience to ignore. To combat this, take a moment before starting to "settle" or "plant" your feet where they are comfortable. Positioning yourself near a table during practice will also help.

A presenter with stellar personal presence effectively combines energy, inflection and eye contact to engage their audience. They use their hands for emphasis but avoid nervous hand gestures and too much body motion. They know their material well without memorizing it.

Exercise 6.4. Identify skills for you to improve; practice to improve them.

- Look back through the skills discussed in this chapter to identify which ones you already do well, and which you need to improve. To double-check, use your phone to videotape your presentation, using the slides on your laptop. Often presenters who have never been videotaped are surprised to see a skill they need to improve.

- Which skills do you already have? Be sure to keep up your performance in these areas.

- Which skills do you need to improve?

  ○ Which did you know you need to improve?

  ○ Which were you surprised to learn you need to improve?

- Describe your plan for improving each of these skills, one at a time, by reviewing the earlier exercises in this chapter or designing a similar one of your own.

As you expand your oral communication skills, practice giving presentations without slides, notes, or handouts. Joining organizations such as Toastmasters will give you experience with this.

CHAPTER 7

Other Oral Communication Skills

In this chapter we will cover nine other types of oral communication skills. We include:

- Challenges in oral communication

- Choosing the right medium for your communication

- Cross-cultural communication

- Listening

- Oral communication by phone

- Oral communication in meetings

- Oral communication in teams

- Using oral communication to build social networks on your job

7.1 CHALLENGES IN ORAL COMMUNICATION

In this chapter, we focus on common challenges faced by engineering or science students, or engineers or scientists, who present. The information is based on over 800 engineering presentations viewed by the author, input from executives, and input from engineering faculty.

The challenges are grouped into three categories: your slides, presenting and interacting with your audience, and your situation. For each challenge we describe the effect on the audience and suggestions for overcoming the challenges.

7.1.1 COMMON CHALLENGES REGARDING YOUR SLIDES

We cover the common challenges of avoiding too much detail and how to connect the details you include to the big picture. And we discuss how to avoid using too complex slides and too many slides.

Many young engineers and scientists have the problem of wanting to share all their data. The key is to provide enough data for your audience to understand, and to connect the details to your big picture or the purpose of your presentation. Here are tips to keep in mind.

Tell the audience only what they need to hear, instead of giving them all the details you want to tell them about.

Use enough detail so your audience will understand your main points, but not so many details your audience will tune out.
Prioritize ahead of time the topics most important for your audience to know.
Regularly remind your audience, at regular and natural intervals, how a particular detail fits into the big picture and why it matters for this topic.
Never just put up a graph and expect the audience to extract your main concept from it. Plan to concisely describe the purpose of your graph and give a concrete example, if possible.

Here are some pointers on avoiding overly complex slides.

Use a reasonable number of words on slides and avoid full sentences. Otherwise you and your audience will tend to read the slide and your audience will stop listening to you. The information should be viewable from the back of the room and readable at a glance.
For charts and graphs, consider, "What is the main point of my chart or graph?" Use the minimum amount of information needed to get your point across.
Avoid "data dumps" or showing all the information to your audience instead of only the relevant information. After identifying the main point of that slide, ask yourself, "What is the smallest amount of data I can use to get my point across?"
Consider using animation. Instead of having information fading in or crystallizing, which can be irritating to your audience, consider having a chunk of the information "pop" in at one time. You may also want to fade the rest of your slide to help your audience focus on your current point.

Figure out the maximum number of slides that will fit within your time slot by practicing with your slides as soon as you have them ready. Ask a person who isn't presenting to time the whole talk and the individual speakers. Consider whether you expect to be interrupted by questions, if so, add two or three minutes to a 15-minute talk. If you and your colleagues have included too many slides, do not "fix" your presentation by deciding to talk very fast. You will need time to discuss the technical part of your presentation slowly enough for the audience to process the information. To decide which slides are less needed, set aside your slides and fill out a storyboard with the main topics that your audience needs to hear. Then return to your slides and objectively (without emotion) remove some or put them in the appendix for answering questions.

When you are in the middle of your presentation, pay attention to warnings you may receive about how much time you have left. Be ready to skip information if you need to, for example, by saying the main point of each slide or skipping one or more slides.

7.1.2 COMMON CHALLENGES OF PRESENTING TO AND INTERACTING WITH YOUR AUDIENCE

In this segment we review common behaviors to avoid, for example, memorization, while presenting to your audience or interacting with your audience. And we describe how to improve other challenges, such as using a pointer.

Avoid Memorization

Avoid memorizing your talk for the following reasons: so you don't lose your place and have to repeat several sentences; so you don't stop and replace words that are already accurate with synonyms; so you don't sound robotic.

Here are suggestions.

Each time you practice, say your presentation in different words.
Do not practice too much—for instance, one student who memorized their talk told me they practiced 15 times. It's very hard to avoid memorizing if you practice that many times. Try three or four times, including the times you time the presentation, and see how that works for you.
Do not write out your talk. Instead, include the necessary but brief cues on your slide and refer quickly to them before turning back to your audience.

Avoid Stressing the Negative

You can be honest yet stress the positive. In our instruction we emphasize "describing the glass as half full, not half empty." In one example from seven years ago, a student presenter described their client's warehouse as "a mess." Instead, we suggested they say "the client's warehouse needs standardization and reorganization."

Student presenters often say, "this is just…" instead of leaving out the word "just." When you use this word, the audience may think you are apologizing for your work. Instead, omit "just." This may take some practice.

Avoid Poor Vocal Quality

The most common vocal issues are: speaking too quickly to be understood, using many "um's," using very long sentences, speaking too softly, and talking in a monotone. Here are suggestions for improving each area.

Speak more slowly. If you are talking too quickly you will see frowns or questioning looks on members of your audience, and they may start looking away from you. In particular, leave time to explain the technical portions of your talk more slowly.
Avoid using "um's." Many of us use some "um's" in everyday conversation, but the challenge occurs when you as a speaker or a meeting member say so many "um's," you distract your audience. Ask yourself what you are thinking when you say "um's." Often people say they're thinking about what they will say next. During practice, then, try replacing "um's" with silent pauses here and there as needed. The pauses will give you time to think and may seem long to you but usually not to your audience.
Avoid using very long sentences. Sometimes when people give a presentation, they use run-on sentences by saying "and…and…and" before they pause. If you do this, practice pausing here and there at a natural break in your sentences so your audience can stay up with you.
Avoid speaking so softly the audience in the back of the room cannot hear you. Speak up, even though you will sound very loud to yourself. In instruction, we would ask you to say after us, "I can be definite!" Then we would ask your team members for feedback while you practice once or twice. This practice is especially important when you are about to talk in a large room.
Use inflection to bring attention to your main points, instead of speaking in a monotone.

Avoid Nervous Actions

Nervousness may express itself in many different behaviors such as your lack of eye contact, moving around so much you distract your audience from your main message, saying many "um's".…. Here are suggestions.

Once you learn you have a nervous behavior, try to figure out what you're thinking as you do it. For example, we discussed "um's" above. In general, consider: "What can I do to deal with the cause of my nervous actions so, even though I am nervous, the audience won't perceive me as nervous?"
Accept the fact that everyone is nervous before giving a presentation—regardless of how much experience they have. Just press on. You may find it helpful to visualize the anxiety as something just sitting on your shoulder—it's there, but it won't stop you from giving a great presentation. You're still the expert on your topic.

Exercise 7.1.1. Ask your colleagues how they control being nervous before and during their inter-action with others in various situations. Describe below what approaches will probably work for you. Then, try the strategies out with colleagues to see if they work.

1. Before a presentation:

2. During a presentation:

3. During a meeting:

4. During a face-to-face interaction with a higher level manager:

Now that we have discussed what to avoid, we review how to do well with two other key communication challenges. First we describe the basics of using a laser pointer, then we suggest strategies for dealing with questions, whether during a meeting or presentation.

Using a Laser Pointer

Many times during a presentation, we have all seen the speaker overuse the laser pointer. Either they leave it on too long, or they spin never-ending circles with it. To use a pointer effectively, practice ahead of time. Decide where you plan to use the pointer—for example, you don't need to use the pointer for every point on your slide, but maybe two or three main points. You may want to use it to highlight two lines on a line graph. Then practice advancing your slides and going back to an earlier slide. Finally, practice pointing to a certain area of the slide and then turning the pointer off. The longer you leave the pointer on, the more likely your hand will shake slightly, which will distract your audience. If you decide to circle or underline something, do it once. Practice will make a huge difference in your effectiveness. Try to practice with the actual pointer you will use.

Answering Questions Effectively

Engineering and science students sometimes rush to answer a question without really hearing it (especially when the question is seen as an objection). Here are suggestions.

Be pro-active: ask for questions between the main chunks of your presentation.
Be open to suggestion: when you do not know the answer to the audience questions, say "I don't know, but I'll find out." Thank the audience member for pointing the idea out to you. Then follow up quickly to respond.

Here are other common behaviors to avoid.

Interrupting the questioner: This behavior shows disrespect to your audience member and can generate a negative attitude. Instead, wait for the questioner to finish before you start to answer.
Answering with a lot of words that don't answer the question: The audience may lose interest. Instead, be brief and to the point. Avoid adding unnecessary information.
Answering the wrong question: The audience member may tell you, either in the middle of or after your answer, "That's not what I asked." To avoid this, if you are not sure you understand the question, ask the audience member to clarify, repeat, or explain a certain part before you answer.
Guessing the answer to the question when you don't know the answer: The audience may be able to tell you are not being forthright—this will decrease your credibility as a speaker. Instead, be honest and direct. If you don't know, say so. You can still end with a positive, for example, "I don't know but thanks for pointing that out. We will look into that." OR "I'm not sure but I will look into it and get back to you with the answer." Then be sure to follow up.
When presenting as part of a team, interrupting other team members as they answer questions: If you add information after a teammate has already answered a question, the whole team will look like it's not functioning well. Instead, before the presentation, decide who will take questions in which areas. And be ready for the case when the teammate expected to answer the question is unsure. In this case, that teammate should check with the others nonverbally, using eye contact. The person ready to answer indicates this with a slight nod, or by putting an index finger in the air.
When presenting to a large audience, answering a question before repeating it more loudly: Most of the audience cannot hear a question asked by a person on the other side of a large room. Repeat the question more loudly so all of the audience can hear it—before you answer it. Otherwise the audience will be trying to figure out what the question was.

Exercise 7.1.2. Which of the above behaviors have you used when answering a question? Write the approximate question down, your behavior, and your plan to improve or change.

1.

2.

3.

7.1.3   COMMON CHALLENGES IN THE SITUATION

In this section we discuss two additional challenges: first, how to present to a mixed audience (technical and non-technical) while keeping the entire audience engaged, and second, how to prepare for situations you don't expect.

Presenting to a Mixed Audience While Keeping the Entire Audience Engaged

Suppose you are presenting to a mixed audience including technical and laypeople. Each time you use a phrase or acronym possibly unknown to part of the audience, define or describe it. Do this whenever you use the phrase so you are not requiring the audience to memorize the meaning. Introduce the descriptions, though, so the more technical audience members realize what you're doing, for example, by saying, "just to make sure we're all on the same page…"

Go the extra step so your audience can easily understand your information. Take the responsibility as the speaker to do the hard work of figuring out how to explain a concept simply.

Preparing for Situations You Don't Expect

Suppose someone challenges what you say. One executive was making a presentation to 100 people and got caught off guard by a heckler. It turned out the heckler was favorable to one of their competitors. After that one experience, the executive goes into a presentation "ultra-prepared" for someone to challenge something they say. So think ahead about possible challenges and how you will respond to them.

In order to avoid unexpected situations:

- If possible, know the specific background and expectations of every audience member before speaking. Then brainstorm about the possible questions (and objections) that may come up.

- Find out details of the setting before presenting; visit the site if possible and practice in the exact setting.

- Expect your technology to fail; be prepared for alternate ways of presenting.

7.2   CHOOSING THE RIGHT MEDIUM FOR SUPPLEMENTING YOUR ORAL COMMUNICATION

Your oral communication will usually be intertwined with another medium, including face-to-face interaction, phone conversations, and electronic communication through E-mail, texting, and instant messaging (IMing). Selecting the medium to support your oral communication can have more effect on the message getting through than the content of the message. For example, the generation or two older than you have their own preferences in communicating. Even though you prefer E-mail or texting, older people may prefer the phone. Check ahead of time to see what medium your audience prefers.

7.2.1   COMMUNICATING FACE TO FACE

Use face-to-face communication if you want to be read the most nonverbal cues possible, and if you are dealing with conflict. This approach also works better if your audience is physically close—such as in a cubicle down the hall. Use face-to-face interaction in these situations, as well:

- if you are challenging the position of an authority above you or on the same level as you,

- if the situation involves large sales,

- when you have too many topics to cover at once in any other medium,

- when asking for favors from people you do not know well,

- when asking for feedback or when giving certain feedback,

- when you are responding to criticism you heard about third-hand,

- when you do not have authority over some individuals but you need to get them to do something,

- when sharing a status report, or

- when the person you're talking with has an effect on the next step taken.

    Be aware of the drawbacks of face-to-face communication:

- You will not have an opportunity to contemplate your next steps, as compared to E-mail communication.

- You will not be able to refer to any notes you may have on your main points, as contrasted with phone conversations.

Finally, here are three suggestions for having effective face-to-face interaction. First, consider making an appointment ahead of time, by phone, E-mail, or text. You will be able to avoid interrupting a person, for example, who may be in the middle of working on a spreadsheet. Second, start your conversation by clarifying why you asked for the interaction. Third, send out a brief summary E-mail after the conversation to confirm the main points of discussion.

7.2.2 COMMUNICATING BY PHONE AND VOICEMAIL

In today's world, face-to-face interaction will not always be an option. Expect to have a great deal of global interaction on your job. In phone conversations, as contrasted with E-mail, you will experience no lapse in your interaction. Using the phone will help clarify these situations:

- When the first question is which topic is most important to discuss.

- When the range of possible conversation is broad. For example, when you expect to find the conversation turning in many ways.

- When you need the back and forth of questions and answers since the answer to one question will lead to the next question.

Be aware, though, if you are a non-native speaker you may find using the phone more difficult than using E-mail or interacting face to face. And, you may have trouble reaching a person by phone. Consider E-mailing ahead of time to request a phone appointment.

Here are suggestions to make your phone calls effective.

Remember you have no subject line, as you do in E-mail.
Use an outline to help you make sure you cover all your main topics.
When you leave a voicemail, keep it brief and speak clearly. Leave your phone number or E-mail address at the end.
When you are about to make a global phone call and this is still a new practice to you, practice dialing internationally (country code before city code) ahead of time. This ensures you will be prompt.

During a call across a great distance, pause when you make a comment, since it will take your colleague(s) about five seconds to respond. And, if you are listening but not speaking, consider using the mute mode to avoid background noise.

7.2.3   COMMUNICATING BY E-MAIL

Using E-mail is convenient for many reasons. First, you can respond to an E-mail at your convenience. Second, your E-mail can capture the thread of your communication. If the thread is long, your contact will benefit if you summarize the main points of the E-mail thread in your introduction while leaving the thread attached for further detail. Third, you can attach electronic documents relevant to the conversation.

E-mail is most effective when you use it in the following situations:

- When you need to reach a person who travels a great deal.

- When you want to distribute information to other people. Once a person receives your E-mail they can choose to agree and pass it on with your information/justification. Or they can question the specifics of your communication and recopy the other people in the conversation with questions or a rebuttal.

- When your message is not urgent and can be read and responded to when the receiver has time.

Of course, you may have already experienced some disadvantages of using E-mail. You may have mistakenly forwarded E-mail to many more people, or a different person, than you intended. This can be a disaster in business. Or, your E-mail's recipient may take a long time to respond. And, miscommunication can happen. For example, in one large business, people regularly E-mailed from their location in Maine to their location in China. One day, a Maine staff member E-mailed their Chinese counterpart, " I am sick and tired of your behavior…" The next day the sender received flowers with a note saying "I am sorry to hear you are sick." Using a phone may have helped avoid this miscommunication.

Every time you send an E-mail, decide who to carbon copy or cc. Since we all get too much E-mail, lots of carbon copying can get unwieldy. Be aware of the following issues:

- Copying entire distribution lists to reach three of four people, just because it is easier for you, is not polite.

- Some people think if they cc more people, somehow they are gaining leverage. Remember no matter who you copy, you are responsible for making sure the colleagues receiving your E-mail understand your point of view.

- Respond only to the sender if they are the only relevant person. For example, if the sender has sent a question like, "Do you still need access to this database?" just reply back to the sender.

To make your E-mail as effective as possible, first, always use spellcheck. Second, use specific subject lines to indicate topic and urgency. If you can fit the message in the subject line, you may be able to save your recipient the need to open your E-mail. For example, your subject line might read, "Are you going to Supplier Review Meeting at 2pm?" Add "(end of message)" or "EOM" at the end. Third, if you don't need a reply, end your message with "No Need to Reply." Fourth, if your E-mail will be going to a large distribution list, or your boss, draft it first and set it aside for re-checking. Remember, once you send it, you cannot take it back. Finally, be very concise, by following these suggestions.

Put the most important idea first, with details below. That is, offer the reader a "net version" (cutting to the heart of the matter) and then offer details or justification.
Include your main ideas on one screen without scrolling, no matter what electronic medium your receiver is using.
Use shorter paragraphs, changing the paragraph for each new thought.
When you have multiple topics to cover, consider sending separate E-mails. For example, send six short E-mails, each with one topic, instead of sending one long E-mail. The shorter E-mails are much easier for your recipient to file or forward.

To end this section on E-mail, avoid the following.

Using "text speak" in E-mail.
Assuming your E-mail reaches only the person you send it to. It can easily be forwarded.
Sending E-mail you don't want the whole world to know. Everything on E-mail is discoverable. It doesn't disappear; instead it stays on your organization's servers, or on Facebook or other internet social networking sites.

Finally, watch out for autocorrect since it can turn words into totally different terms. In one example of a 'communication fail," the word "sure" was changed to "sire."

7.2.4   COMMUNICATING THROUGH TEXTING

Texting is best used when your messages are very short and when your receiver understands the abbreviations and vocabulary used. Sometimes people use texting only for their own personal communication. In business, check the expectations of your boss and colleagues. Although generally texting tends to get a quicker response than E-mail, remember that people in different generations may not know the abbreviations. Check with them in advance.

7.2.5   COMMUNICATING THROUGH INSTANT MESSAGING (IMING)

IM is best used when you think you can accomplish the communication faster, when you know the other person is in a meeting, and when the matter is urgent. IM is helpful on webinars or multi-person video conferences. It does not interrupt the flow, yet can guide the dialogue. Also, IM can be kept private for people on the same side of the discussion, for example, when Company A is selling to Company B. The main drawback to IM is when the person receiving the message does not know the abbreviations and vocabulary.

Balance the benefit of delivering that essential information against the intrusion into the time of the person you are messaging. As a rule of thumb, a person should be able to read an IM in a few seconds and respond in an even shorter time. If at all possible, phrase the question so the person can answer "yes" or "no."

Use IM when:

- you think you can accomplish the communication faster, and the situation is urgent,

- you know the other person is in a meeting,

- you know the receiver knows the abbreviations and vocabulary, and

- you need to reach multiple people at once. For example, you can have five or seven conversations going at once.

7.3   CROSS-CULTURAL COMMUNICATION

You will need excellent cross-cultural communication skills while doing business with people of other cultures—whether the culture reflects a different country, one part of a country, or a difference in age between two groups of people. First, allow enough time for preparation ahead of time for cross-cultural communication, and second, focus on knowing what you need to avoid. For example, when an American is presenting building plans to an audience in China, they need to know that a main concern of the Chinese is "Feng Shui," a system of rules or laws governing spatial arrangement and relationships relating to the flow of energy. In the West, this concept is often not taken seriously, but the approach is very serious to the Chinese.

In general, when interacting across cultures, understand the culture you're speaking to. To help you prepare, this chapter covers the following:

- Demonstrating respect, humility, and a willingness to understand

- Learning the following types of information as part of your preparation:

  ○ Learning what to avoid

  ○ Knowing the local traditions and mannerisms, such as where to sit in a meeting

- ○ Learning at least some of the language and what defines proper or expected dress

- Working with translators and "local champions"

- Interacting with others whose culture is different because they live in one part of a country, or because of age.

7.3.1 DEMONSTRATE RESPECT, HUMILITY, AND A WILLINGNESS TO UNDERSTAND

To show respect, be prepared and open. Learn the information, described here, necessary for successful interaction.

Avoid arrogance; never assume you are better than people from another culture. Instead, be humble.

Be open to new ideas and your audience's values.

Learn more about the differences in your and the other's culture as you continue to interact. For example, when presenting, pause more frequently and ask if people understand the point you just made.

7.3.2 LEARN THE FOLLOWING TYPE OF INFORMATION AS PART OF YOUR PREPARATION

Learn what to avoid. One executive mentioned that they couldn't count the times they were sitting in a room with Europeans or Asians listening to an American speaker using analogies or expressions that are unrecognizable to people of other cultures. The executive could clearly see the expressions of confusion on the Europeans' or Asians' faces. Do not use jargon or "Western slang." Examples of slang include "way too much," "water under the bridge," "barking up the wrong tree," and many others.

Learn the local traditions and mannerisms. Know what is expected to happen before the meeting. In some cultures, two or three purely social visits are required before any business is discussed. And be aware of how to meet a person: Do you bow? Do you shake hands? How do you address the person (what title should you use? Are you expected to use their first or last name?) How do you present your business card?

Learn the expectations that apply during a meeting. Know the answers to the following questions:

- Is it acceptable to use digital equipment such as cell phones during the meeting?

- What does the seating mean and where should you sit? In some cultures, people with more power sit in the middle of the table instead of at the ends.

- What sort of supportive materials are you expected to have? Electronic or hard copies or both?

- What do various verbal responses mean? For example, a verbal "yes" may mean "no."

- What do different nonverbal behaviors, such as eye contact or lack of eye contact, mean?

- Are you expected to talk less and spend more time listening?

- What ways are used to confirm understanding? For example, instead of asking "did you understand this last point?" you may need to ask more specific questions or repeat ideas in several different ways.

- How will the meeting notes be handled?

- How will the meeting end?

- What will the follow-up look like? Who will handle it? Are you expected to send a thank you letter? A summary?

Learn as much of the language as possible to show respect. For example, know how to greet people and how to say good-bye. Know at least some basic words, such as "yes," "no," or "we have an agreement."

7.3.3   WORKING WITH TRANSLATORS AND "LOCAL CHAMPIONS"

A translator does their best to convert your words into the other language, but sometimes they miss the nuances based on your nonverbal gestures or facial expressions. To avoid this, find and work with a "local champion," or a person working in that company who supports your agenda and understands English well. Ask the person to help you check that the translator gets across your real meaning.

7.3.4   WORKING WITH PEOPLE IN CULTURES BASED ON ONE PART OF A COUNTRY OR BASED ON DIFFERENT AGE GROUPINGS

Sometimes other cultures are defined by part of a country. For example, in the U.S., people from the Northeast sometimes speak faster than people from the South. And in the South, the use of "you all" in place of "you" does not necessarily indicate informality. Learn and allow for local or regional mannerisms or values.

Sometimes other cultures are based on different age groupings. For example, one executive said many people who report to them are much older than they are. The subordinates have also been at the company longer than the executive. When working with older people, the executive suggested that you:

- listen carefully with an open mind.

- Be ready for some older people to talk more slowly. Do not take this to indicate lack of intelligence.

- Emphasize working *with* these people instead of stressing that they are working for you. Hierarchy is very important in some cultures, so know when and where to use this approach.

- When working with an older person with more experience than you in their setting, avoid telling them what to do. You will not succeed. Instead figure out what contribution you can provide as their director and how you can connect with them.

Exercise 7.3.1. Working with people of different ages.

Imagine yourself working with a start-up just out of college. You may be in Silicon Valley, Boulder, Colorado, or Berlin, Germany—one of the fastest-growing start-up communities. You are one of the many young people working at the fledgling company, while several key people are older and have lots of experience in solving certain kinds of problems.

Identify key points you would keep in mind when interacting with one of the older people.

1.

2.

3.

4.

7.4   LISTENING

Listening is an essential part of oral communication that is often overlooked. It is the skill of "paying attention." In this chapter we cover executives' suggestions on:

- Situations where your audience has an opportunity to tell you what they expect

- General listening tips

- Listening in specific types of situations

- Listening in two special situations:

  ○ How to listen when you are the final decision-maker in a group discussion

  ○ How to listen when you're part of an emotional discussion, for example, a talk with an employee about job security

7.4.1   HAVING THE OPPORTUNITY TO LISTEN TO WHAT YOUR AUDIENCE EXPECTS

Whenever possible, get feedback from your audience about their expectations before you present. This applies to formal presentations as well as less formal one-on-one updates. Once this feedback is given, tailor your style for the most effective conversation. As the presenter, remember that, although these comments may seem very direct, their purpose is to help you communicate more effectively with the audience. Here are examples of executive comments to presenters before their talk.

"I have five minutes to get the main points so be net." (or get to the heart of the matter quickly)
"I want to understand this from top to bottom, so if we run over our allotted time, I am OK with that. I am not leaving this conversation until I understand the issue."
"Assume I know nothing about this topic and I'll tell you when you can skip ahead."
If you are giving too much detail, the executive in the audience may say they appreciate your depth of knowledge, but ask you to "hold that for later until I get the framework digested. We can then decide if it warrants coming back."

General Listening Tips

In general, show you are listening by using plenty of eye contact. While you are listening, respond with "um-hmm" or a nod of your head to let the speaker know you're following them. If the situation permits, interrupt the speaker with pertinent questions for clarification—at the time of your question. This avoids the necessity of backing up. If you are not sure if you understood something correctly, ask to reflect back to the speaker what you think they just said. Their response will tell you whether you have understood.

Do not rush the speaker, and keep your attention on them without letting your eyes drift away or taking cell phone calls. Show you are open to ideas by using open gestures such as holding your palms up.

7.4.2 LISTENING GUIDELINES FOR CERTAIN TYPES OF SITUATIONS

Different situations require various types of listening skills. Here are seven different contexts and the suggested listening mode.

Situation	Listening Mode
A person or group seeking your advice	Listen to the person or group seeking advice with constant questioning to ensure you understand the details and complexities of their situation
A brainstorming session	Listen but also build on what others say, always in a positive way
A debate where both parties are going in knowing they disagree	Listen to the facts and arguments without emotion in order to respond directly. Do not use prepared thoughts
An emotional dumping, complaining, or venting session	Listen without judgment or qualifications
A casual getting-to-know-you conversation	Listen and ask questions to show your interest in more than just the business topic—for example, family life
A group discussion, especially when spectators are present but not contributing	Listen to make sure all discussion is on point and appropriate for the audience
A discussion involving people who are subordinates, peers, more senior than you, or a mixture of all these types	Listen for clues as to who is the decision maker and who is the subject matter expert; who is a dissenter and who is a supporter

7.4.3 LISTENING TIPS FOR TWO SPECIAL SITUATIONS

Here are suggestions on being an excellent listener in two special situations: first, when you are the final decision-maker in a group discussion; second, when you are part of an emotional discussion, for instance, a talk with an employee about job security.

How to Listen When You are the Final Decision-Maker

If you are the final arbiter, listen actively to the group discussion. Pull out the ideas expressed from each group member to create an environment where free-form thinking is welcome. Be sure you are listening and actively seeking feedback from all parties (including the shy ones) before giving your opinion. One executive said, "In a team of subordinates or even peers, if they know the call is ultimately yours, hold back waiting [to give...] your signal as to where you stand on an issue. Don't wade in too early [with your own opinions]."

How to Listen When You are Part of an Emotional Discussion

Whether you are the employee or a manager in an emotional discussion, take the time to think about what the other person really means. If you are a manager talking with an employee during a performance review, or talking with an employee about job security, here are some guidelines.

In emotional discussions, often the listener will want to solve the problem so the speaker feels better. Fight this tendency; let the person speak their mind completely, exhausting all their thoughts—even if this means suffering through some awkward silences.
"Write everything down when you're in an emotional state." It may be helpful to first ask if it is OK and give your reasons. This will let the other know you do not have a sinister motivation, but instead you are appropriately serious.
One executive recommended offering supporting statements but: "be brief, don't take over the conversation, and don't launch into a story about your own experiences...unless you are absolutely certain" that the person wants to hear this.

As a manager, engineer, or scientist or engineering or science student, understand that some people often make exaggerated statements in an emotional discussion. If a person typically does this and tells you, for example, "you're going to fail," interpret this based on the fact that this person often makes exaggerated claims.

Listening, as with other oral communication skills, improves with practice. Being aware that you can improve your listening skill is the first step.

7.5   ORAL COMMUNICATION BY PHONE

Oral communication by phone is a large part of the engineering or science student's and professional's work day. The following issues are central to making effective professional calls: Deciding whether to call or use another medium (such as E-mail or person-to-person interaction) to communicate, preparing for a phone call, making the call, and following up after the call.

7.5.1 DECIDING WHETHER TO CALL

Identify the other person's preferred mode of communication, and use it. If it is E-mail, for example, then use that medium unless there's a request for a call.

Here are guidelines to help you decide whether a phone call will be more effective than E-mail or person-to-person interaction, for example.

A phone call is more effective than E-mail if a discussion needs to take place for a decision to be made. Once one person has added information, the other needs to respond, and so on. A phone call is more effective since there is no lag between comments.
A phone call with E-mail is more effective if you haven't had any previous contact with the person you're calling. Contact the person with E-mail, tell them the topic and questions, and ask for a good time to call. Follow up with a phone call to ask and answer questions.
Phoning is more effective than personal interaction when the person is geographically distant from you, when video conferencing or Skype is available, and when travel costs are not available.

There are cases when person-to-person interaction might be more effective than a phone call. One case is when the person is located geographically very close to you. One executive said they had to keep reminding their consultants to interact in person when their cubicles were close to each other. A second case is when you have regular business interaction with another person and you haven't seen them in a long time.

Exercise 7.5.1. List three variables that come into play when deciding whether to call someone.

1.

2.

3.

7.5.2 PREPARING FOR A PHONE CALL

Many people just pick up the phone and call. Planning your call will greatly increase your success. Keep this in mind:

Know your audience. For example, gather information about them from a variety of sources, especially if you don't know each other well. Ask yourself:

- How will the person benefit from our phone interaction?

- What are the best times of day to reach this person by phone? For example, executives are most likely to be available for a call at the very beginning or at the very end of their business day.

- What experience and background do you know they have...so you can shorten your interaction?

- Is this time of year an especially busy time for them?

Focus on being as efficient as possible. Respect the other person's time. One executive noted if you ask for five minutes and take 15, "it is likely to backfire." Ending your call on time is especially important if you want the person to welcome your future calls. Then, stay focused on your topic and goal. Include only necessary information. One executive suggested if the call is really important, consider writing out a script or having an outline to speak from. Finally, clarify your purpose and keep reminding yourself of it. *Be clear about what you want to have happened at the end of the call.* Here are some possible purposes:

- **Giving an update.** Asking for information or gathering information, such as when a Capstone Design student group calls their client to find out exactly who will attend their next presentation, who may attend, and what their backgrounds and expectations will be.

- **Asking for help or assistance**.

- **Asking for a meeting.** One executive described a call where the caller said their purpose was to "lock down" a time for a meeting they wanted the executive to attend. Instead of providing a very brief description of the topic and discussing possible times, the caller enthusiastically started describing details about the topics to be discussed. The call was like an unprepared elevator pitch. As a result, the executive started deciding what they felt about the topic right then, before any discussion, and they considered not even attending the meeting.

- **Be sensitive to the needs of the person's daily schedule.** If something urgent comes up, be ready to reschedule.

- **Be confident, but respectful.**

- If the person you're talking with tells you about examples from their experience:

- ∘ If appropriate, briefly describe examples in your experience at the same level of detail

- ∘ Be ready, if appropriate and if the information is not confidential, to briefly describe an example from another conversation with someone at their power level (for example, maybe a C-level individual…CEO, CTO). They may be more open to talking with you because you have experience in talking with individuals of their level.

By doing these things, you are building what you have in common and creating a networking connection for yourself.

Exercise 7.5.2. How long are your calls? Time some if you're not sure. Now, what can you do to shorten them?

Call number	Date	Length of call	How to shorten the call next time
Call 1			
Call 2			
Call 3			
Call 4			
Call 5			
		Avg. length=	

Now, apply your ideas for a week. Check how long your calls are now.

Call number	Date	Length of call
Call 1		
Call 2		
Call 3		
Call 4		
Call 5		
		Avg. length=

Repeat as needed to shorten your calls.

Use inflection in your own voice to keep the other's interest and listen very carefully to their voice since you will not have other nonverbal cues. A good listener will check for the following:

- **Listen for a distracted tone.** Are they thinking about something else? It may be best to continue the conversation another time.

- **Is there any hesitation on their part?** If so, ask about it. They may be changing their mind, or they may be thinking about how to answer.

- **Are they talking very fast?** Fast speaking may indicate they are under pressure for another deadline, or they may be indicating your time with them is up. Always assume your request is not their first priority, and ask when to call back.

Exercise 7.5.3. How good a listener are you on your phone calls? For your next three or four phone calls, check which of the three things above you did on the call. How can you improve?

Call number	Date	Did you listen for receiver's distracted tone?	If receiver shows hesitation, did you ask about it?	If receiver is talking fast, did you ask to reschedule the call?	How to improve
Call 1					
Call 2					
Call 3					
Call 4					

7.5.3   MAKING THE PHONE CALL

Phone calls in different contexts require different approaches. Here are tips on all calls, including *cold calls*—when you do not know the person you're trying to contact, and on your discussion once you reach your targeted person.

For all phone calls, be confident but not overbearing, have respect, and expect cooperation. But if the person indicates they cannot talk with you or respond to your request, thank them and sign off quickly. Use your outline as "talking notes" and take notes on the conversation. Listen for voice cues.

For cold calls, ask for the person's assistant and ask for their help. Mention your referral if you have one, and then identify clearly and concisely what you need. End by asking for advice on what to do next (for instance, the assistant may ask you to send a brief E-mail) and follow the exact procedure suggested.

When you reach the person you want, whether after an initial cold call or on your first try, ask your target person for help. Ask if this is a good time to talk for a few minutes; if it is, keep track of your starting time and required ending time. Identify, clearly and concisely, what you need from the person. If the person agrees they can help, identify next steps and ask if they agree. Finally, if they do agree, thank them for their time, sign off, and make sure you take the next step right away.

Of course, your particular call will vary depending on its content. Many of these guidelines will apply to every call.

7.5.4   USING SKYPE OR VIDEO CONFERENCING

If your phone call is done using Skype or a teleconferencing service, give special consideration to these topics. This type of oral communication will increase in the future.

Using Skype

If your phone communication is via Skype, follow these guidelines.

Do a technical test first, whether one hour before or one day before, with enough time to fix any possible problems. For example, you may encounter a delay in the voice, or the connection may drop so you can see the people but not hear them. If Skype doesn't work during the actual call, one executive advised rescheduling rather than trying repeatedly to make it work.
Call from a professional environment. Be aware of what is behind you. Be sure you (and the person on the other end) will not be distracted by interruptions or noise.
You will not be able to change the lighting. If too much light is behind you, you may look like a ghost. This will depend, of course, on the quality of your computer and your camera.
Be careful of your movement. Avoid crossing your legs during the call, for instance, since your shoulders will move in your Skype picture and your audience may interpret your movement as evidence of your loss of interest in the conversation. Have your hands on the table so you can use them here or there if you want, and lean 45 degrees into the screen to show your interest in the conversation. Sit close to the screen to be seen well.

As you start the conversation, ask "can you hear me well?" and "can you see me well?" and make any needed adjustments.
If the sound stops working, you can use the phone along with Skype, but it will be a distraction both to you and your audience.

Using a Teleconference, or Video Conference, Service

Here are suggestions from executives on what to do and what to avoid when you're communicating via a Video Conference (or VC) Service. In one common situation, you may be asked the zip code of your location, and then you will receive driving directions to the closest center. About 15 minutes before the call, the technician will set up your technical tools. Here are guidelines.

You have control of the zoom, the light, and the volume.
Be seen in the frame of the screen just as if you were across the table. Look at yourself and zoom in, then look at the other person. Make sure the zoom is right at eye level.
As with Skype, when you start the conversation, ask "can you hear me well?" and "can you see me well?" and make any needed adjustments.
Focus on what you say, not what you're doing. In this type of conferencing, vocal characteristics and eye contact play a much larger role than your physical presence.
Still, eye contact and posture are very important. Lean forward by 45 degrees to display your engagement, keeping your hands on or near the table.
Do not read what you will say; this will be seen as disrespectful.
Avoid distractions, since your audience is looking only at the screen. In an in-person meeting, people may look at other things in the room for a moment here or there, but here your audience is always looking only at you.

It is possible to have a group discussion on either side of this type of the communication, but people tend to talk at the same time.

There are also combinations of calls using different approaches, such as a conference call when a group is connecting via Skype and another person is calling in by phone. In this case, be sure to pay equal attention to the people you can see and the people you cannot see. Recently a student told me their group had a much harder time remembering to interact with the person on the phone in such a scenario. These types of phone calls—including software of some kind—are expected to proliferate in the future.

7.5.5 AFTER THE PHONE CALL

After your call, follow up by sending thanks in an appropriate way…either by E-mail or, as some executives prefer, by snail mail. Follow up quickly on the promises you made. Consider making sure you and your targeted person are on the same page by E-mailing the main points discussed and what was decided. Start the E-mail with a phrase like, "Thanks for your time. Just to make sure we're on the same page…" or "I just wanted to confirm with you several things we discussed by phone."

7.6 ORAL COMMUNICATION IN MEETINGS

In this chapter we cover executives' input on stellar oral communication in meetings. For both meeting attendees and meeting leaders, we describe the "do's and don'ts" of meeting communication. We review when to speak up and when not to speak up, and the pervasive use of cell phone/blackberries during meetings.

7.6.1 GUIDELINES FOR YOU AS A MEETING ATTENDEE

Here are the actions of a successful meeting attendee: before, during, and after the meeting. We also cover what actions to avoid.

Before the Meeting

As a meeting attendee, "complete your homework" before the meeting. Here are steps to take.

Note who was invited to sense what the dynamics of the meeting might be.
Know why you've been invited.
Review the agenda and relevant "pre-reads" or draft presentations. Note any questions you want to get answered during the meeting.

Check other background information, such as:

- data or information needed to answer questions that may arise

- web information on the other participants

- action items from earlier meeting minutes

- commitments you made at prior meetings

Decide if you want to offer to take notes and publish them after the meetings. This was highly recommended by many executives. One said, "When I see someone taking notes and then distributing them after the meeting I really value their initiative." Your personal power is enhanced if you become the meeting scribe.

During the Meeting

Here are actions to take during the meeting.

Quickly identify whether you're in a "working meeting" or an "informational meeting." In working meetings people seek and encourage discussion; informational meetings are primarily for information dissemination. Don't try to turn an informational meeting into a working meeting. If you have suggestions, cover it with the leader afterward—not in a room full of people who want the meeting to be as short as possible.

Pay attention to what is being said; focus on and respect the people speaking. If you look distracted the speaker may take this attitude as disrespect.

Take note of who sits where and who sits next to one another; sometimes you'll get insights into behind-the-scenes friendships and working relationships.

Participate and engage where appropriate. Here are guidelines on speaking up:

- If you don't understand, ask a clarifying question.

- Speak up if information presented contradicts or is inconsistent with information presented earlier.

- Speak up if you are asked a direct question. Have data that supports your answer. Be clear, direct, honest, and concise. Use a "half full" approach, as opposed to a "half empty" approach. This is more positive but still factual. For example, instead of saying "Their program could not work with this," you might say "Here is what we did to make sure their program ran."

- Speak up if you have a good idea and know how it might work, and if the idea is relevant to the meeting discussion and most of the participants. (Otherwise, speak with the appropriate people after the meeting.)

- Use the time just before and just after the meeting and during breaks to meet the people you don't know.

- At the end of the meeting, if you are not sure of your assigned task and deadline, ask for more information.

Exercise 7.6.1. Speaking up at meetings.

1. Did you speak up in your last meeting? Yes ___ No ___

2. If you did not, what was your reason?

3. If you did not, when was the last time you did speak up at a meeting?

4. What was the purpose of the meeting where you spoke up, and who attended the meeting?

5. Which of the six reasons described in the table above describes why you spoke up? Was it for more than one reason? Yes ___ No ___ If yes, describe the other reasons.

After the Meeting

After the meeting, if you have offered to be the scribe, produce the minutes in a timely way. Summarize the key points and supporting evidence. Consider asking the meeting leader to review the minutes in case they want to add something. Then circulate them as recommended by the leader.

Pay close attention to the activities you committed to during the meeting. Follow up quickly and make sure your information is accurate. Remember, if you are new to this set of meetings and group of people, first impressions last.

Actions to Avoid in Meetings

Now that we've discussed suggestions for effective meeting communication, we'll focus on the behaviors meeting attendees should avoid. Here are some tips.

Avoid speaking up if:
• you are going to say why something will not work without an alternate idea about how it might work. • your new boss asks you not to speak up in the meeting. Before a meeting, discuss expectations with your supervisor. One executive told a new hire not to speak up in a meeting they were both about to attend. The new hire did speak up, irritating the boss. As a result their group was committed to doing something the boss didn't want to do and the boss had to work to change that expectation.
Avoid criticizing other groups or people.
Do not interrupt.
Do not talk about how the topic will impact you personally.

Do not constantly "fiddle" with your mobile device. If at all possible, put it away until after the meeting. If you must check it regularly, do so quickly, respond quickly, and then focus back on the meeting.

Exercise 7.6.2. Explaining reactions of other attendees.

There are specific reasons to avoid the above behaviors. One reason is the reaction you may get from other attendees.

- How might others react if you criticize them in a meeting?

- Have you even been in a meeting where one person constantly interrupts the others? Yes ___ No ___. If you have, how did the others react? If you haven't been in this situation, how do you think the other attendees would react?

- What do you think meeting attendees would do if one person points out the ramifications, on a personal level instead of a business level, of the action being considered?

7.6.2 GUIDELINES FOR YOU AS A MEETING LEADER

Meeting leaders' oral communication can enhance the meeting's success. To be an effective meeting leader, consider the following guidelines.

For a meeting to be positive, you should establish an environment of openness, transparency, and interest. They should also encourage a high degree of interactivity.
Know the people in the meeting going in. And know them in the meeting, that is, read them and pay attention to them.
If the attendees are not paying attention or if multiple conversations get started in the meeting room, ask a question to get the attendees involved again. For example, you might ask if they are already familiar with the information you're covering, so you can skip ahead. Or you might want to say "we seem to be losing focus…can I ask that we bring our attention back to the conversation at hand…"

> Keep it simple. Focus on getting three, or possibly four, points across. You may want to write the three things down before the meeting.

7.6.3 THE PERVASIVE USE OF CELL PHONES/IPHONES/LAPTOPS

During most meetings, participants are multi-tasking. Often you cannot tell whether their use of cell phones, iPhones, iPads, or laptops is supporting the meeting or distracting them. For example, when two people from one company are meeting with people outside their company, they often text a quick message to their colleague suggesting questions to bring up.

Executives noted different approaches to the use of electronics during meetings. One suggested, "If you're concerned about blackberry and cell use during a meeting, address it up front. Ask everyone to turn them off if it is a critical meeting." They added that this was tough to do if you are a junior person. So if you are relatively new, you might want to ask a senior person to request this. Do not let the electronics dominate the meeting environment. Brainstorm ahead of time with your colleagues and supervisor regarding how you would handle this situation.

7.7 ORAL COMMUNICATION IN TEAMS

In our discussion of team communication, we cover its importance and general guidelines for team members and leaders.

Whether you are a staff or line person, communicating in teams is incredibly important—executives note: you cannot opt out of it. Learning to communicate in teams is a vital part of any scientist's or engineers' education. How you communicate is more important than your technical skills. If you don't master team communication skills, your career can be de-railed.

7.7.1 GENERAL GUIDELINES FOR TEAM COMMUNICATION

In this section, we describe, for team members and leaders, how to be aware of your audience, different ways of building your team, and how to collaborate and keep lines of communication open.

If you know your audience, you will know who will have questions and what kinds of questions. Be careful about how people take your comments, so "when you think you're throwing a pebble [you won't] find out it's a boulder." This situation occurs when you haven't done your homework. Focus on being courteous, caring, and genuine in the way you ask questions.

In order to build your team, spell out or say what makes this team strong. Find out who has which areas of expertise.

Always do your best to collaborate. No one, individually, is better than the team. The diversity of thought processes and approaches results in the best final product. As you get more and more responsibility, you cannot do it all yourself. Collaborate to get wisdom, not necessarily consensus. Remember to be open to suggestions. Your colleagues have great ideas, too, so strategize with them about making their ideas work.

Keep your lines of communication open in two ways:

1. Treat everyone with respect. Encourage interaction among your group members by pointing out "this is serious business," but you can have fun while you're doing it. Explain that you do not mean inappropriate fun, but lightness and the ability to laugh at yourself as you discuss serious topics.

2. Open communication must happen in more than one channel. To keep people up to date, use a combination of channels such as conference calls, internal live television (live broadcast with open phone lines at certain points), chats, newsletters, and leadership-driven meetings. Using multiple channels makes it easy for people to speak up.

7.7.2 COMMUNICATING AS A TEAM MEMBER

Sometimes you will have some input in who else is on your team. But more often, a team is thrown together to get something done. When this happens, you may not like some of your team members. Especially in this case, the balance of speaking versus listening is important. There is real value in listening twice as much as you speak. Then, be assertive but not arrogant by saying what you think and want. Avoid being aggressive and offending meeting attendees by using phrases such as "From my point of view…" or "From cases I've seen…"

Here are guidelines for successfully speaking up in meetings.

Keep comments fact-based.
Pay attention. Give the meeting your full attention by not using your electronics unless necessary. If you have to check in, do it quickly and quietly and then focus right back on the meeting.
Be punctual with communication. Be respectful of others' time. Punctuality is good manners but it is also a subtle demonstration of respect. If someone asks for input by a certain time, make the deadline.

Don't be afraid to argue in team settings. Do it respectfully, but advocate your point of view. Then, don't be afraid to change sides if someone has a better approach to the issue. Learn how to qualify your opinions, for example, by saying, "Under normal conditions when the interest rates are high, it would be best if we don't have inventory." Or, "Yes, but if you're facing a shortage of materials…"

Below are actions to avoid when interacting in a meeting.

Avoid showing off. Don't always have to be the smartest person in the room.
Avoid name-calling. Keep the discussion impersonal by focusing on the product or deliverable instead of one individual versus another.
Do not be passive-aggressive, for example, sitting in the corner, tuning out the discussion, and acting like you don't care. This is worse than being actively aggressive. If you do need to be aggressive, don't ever make it personal. Even if the other person starts it, don't follow up.
Don't "spin." Sometimes people do this to show they can do something fast. People tend to do this when interacting with more senior people. Don't be unrealistic about commitment, instead use candor and have a can-do attitude while being realistic. This approach enhances your credibility and it also helps senior people understand how difficult your task is.

When using E-mail for team communication, keep it concise. If you're using an attachment, tell your audience what to notice in the attachment, particularly if it's a spreadsheet. Reinforce the main points of the attachment in your E-mail, since some people won't have time to click on the attachment.

7.7.3   COMMUNICATING AS A TEAM LEADER

Two kinds of team leaders exist: 1) a project manager, or a person who manages the project budget and communication, who may not be an expert in the field; and 2) a subject matter expert, who often leads the team with the project manager supporting them. Here are suggestions for both types of team leaders.

Encourage the team members to participate. Even if you thought you knew each person's expertise, you will see "gems" you didn't know about. Ask questions of people not participating. Be inclusive, and look for any nonverbal signs that a person wants to speak up but is hesitating. One executive noted that, as team leader, you will be building trust along the way: "Trust, candor, and approachability unlock the greatness of a team and result in better insights and better followership once a decision is made."
Present discussion points and arguments as objective problems rather than one person against the other. De-personalize the issues.

Respect the attendees' time by keeping the team meeting conversation focused. Generate respect and trust by being present mentally.
Use formal tools like project plans, agendas, and meeting note formats, to save the team members' time.
Summarize, at the end of each meeting, which team member has taken on which tasks, and the due dates for those tasks.
Set a rhythm for team updates. Be regular, consistent, and persistent about communicating team status.

Exercise 7.7.1. Learning more about being a team leader.

During your next two team meetings, carefully watch the proceedings while you participate.

1. Which of the above guidelines were put into practice by each team leader?

Team leader in meeting 1:

Team leader in meeting 2:

2. Did the two team leaders use different approaches than the ones described above? Yes ___ No ___ If yes, which team leader seemed more effective to you? Explain why.

7.8 USING ORAL COMMUNICATION TO BUILD SOCIAL NETWORKS ON THE JOB[1]

This chapter includes pointers on using oral communication to build your network when you are in a new job. Having high-quality networks enhances your learning and performance, and your job satisfaction. We cover the goals at the beginning of the process, helpful steps, and judging the networking support available in a company where you are interviewing.

---

[1] This chapter is based upon the work of Dr. Russell Korte, Colorado State University. He interviewed newly hired engineers and their managers. Further resources are given at the end of this book. The "boxes" of information are directly from executives interviewed by Dr. Norback.

7.8.1 IMPORTANT GOALS WHEN YOU START TO FORM YOUR NETWORK

At the very beginning of forming a network, be aware of the four goals described below.

First, find a mentor. As you start your new job, think about who you could ask to guide or mentor you. Consider developing a small team of mentors to help—different people have helpful advice in different areas.

> Start finding out about projects where the team is being formed, or look into which team may need more help. Ask your boss if there are projects you can be observing or meetings you can be attending. As a new meeting attendee, be aware that you may be asked to make copies or take minutes in the beginning. During the meeting, identify which people are being delegated tasks. After the meeting, ask to see them to get information on how to distinguish yourself.

Second, start building relationships long before you might need help from others. Remember, relationships go both ways. Consider the value you can offer to others.

> Take the initiative to talk with people around you. Find out how your responsibilities connect with the other parts of the company.

Third, become known by others in your organization. Focus on increasing the number of others in your workplace who know you and the skills and attitudes you bring to the table.

> Make yourself visible and available. Ask questions, and listen. Check the corporate calendar and the intranet and go to meetings that are open. If you are interested in working on a project, ask, for example, "May I participate in this project in this capacity?"

Become a valuable member of your work group. To obtain this goal:

> Observe the people who are distinguished—who have too many projects. Then become someone they can rely on.

Exercise 7.8.1. Finding a mentor, or finding a second mentor.

Plan action steps to identify a mentor, or more than one mentor. Start by answering these questions.

1. Have you found a mentor yet? Yes ___ No ___ If yes, how has this mentor helped you?

2. How can you be of value to your mentor?

3. Are there other areas in which you would like a mentor? Yes ___ No ___ If yes, list these areas.

4. Identify the steps you will take to find another mentor. Are potential mentors:

   ○ In meetings? Which ones?

   ○ In group meetings or one-on-one meetings, or both? Identify them.

   ○ In other settings, such as the social interaction outside your workplace? Which other settings?

5. When you ask a person to be your mentor, be specific about what you'd like them to help you with.

7.8.2 STEPS TO TAKE TOWARD YOUR NETWORKING GOALS

Collaborate. As a newcomer, spend time to regularly consult with others in your work group about your work tasks. Collaborate with the others on completing the tasks.

Build relationships through social interaction outside of work and get to know others personally. As a new employee, focus on getting to know your work group through lunches, sports, or other gatherings. To build these relationships, expect to share information about your personal interests or family activities.

Contribute beyond the expectations of your job function. Help with or complete tasks outside of or beyond what your work group expects of you.

Learn "how things work" in your new job by:

- Keeping your group members informed of the status of your work tasks. Notify them in advance of any particular issues or problems that may arise.

- Exchange information, in group meetings, that will enhance your group's productivity.

- Learn your new job's vocabulary. Examples include acronyms, abbreviations, and jargon, such as the term "apron" for employees on the floor in a major home improvement company.

- Come to know how to resolve conflicting ideas and information with others in your work group. For instance, during disagreements, speak without emotion about how to resolve the issue, and respect the other's point of view.

- Learn how to present to and persuade your work group members. Describe your point precisely and ask for others' input. And learn how much each group member values certain results.

7.8.3   HOW TO JUDGE THE NETWORKING SUPPORT AVAILABLE IN A COMPANY, WHILE YOU INTERVIEW

When you apply for a job, your interviewing phase will probably involve meeting with a Human Resources person, then your potential boss, and possibly several work group members. Use these meetings to gather information about the support provided by the company. Consider the following steps.

Ask the appropriate person, in your judgment, whether a mentor will be assigned to provide regular and constructive feedback for you.

1. Ask whether the company expects other work group members to help new hires become part of their group.

2. Discuss the opportunities, encouraged by the organization, for social interaction outside of work—such as dinners, lunches, volunteer work, and sports activities.

CHAPTER 8

Advanced Oral Communication Skills

In this chapter we cover two more complex oral communication skills. In the first part we describe how to make an effective elevator talk, in case you run into a CEO or high-level person in your organization. Then we describe how to make the most of meeting a person on an airplane. Finally, we describe how to create and present an effective poster.

8.1 MAKING EFFECTIVE ELEVATOR TALKS

Elevator pitches or elevator talks are very brief conversations you may have when you find yourself in an elevator with a C-level manager from your company—whether it is your Chief Executive Officer (CEO), Chief Technology Officer (CTO), Chief Information Officer (CIO), or Chief Financial officer (CFO). These types of talks are often useful in a second setting: when you are sitting next to someone you don't know on an airplane. We discuss the first setting and then the differences in the second setting.

In an elevator, you have an opportunity to get noticed and get input from your C-level manager, but you only have so many floors (7? 10?) to make your point. To prepare yourself for this kind of opportunity:

- Become familiar with the leaders in your organization. Know what they look like, their job title, and what they care about. For instance, the CEO is often interested in the bottom line or information about competitors. The CIO or CTO care about technological issues, while the CFO is interested in financial information.

- Identify what you want to happen as a result of your elevator pitch. Your goal may be to continue your conversation, for instance, to talk with your C-level manager outside the elevator.

- Clarify the main point(s) you want to cover. Focusing on one topic is best, especially if the information will have a big impact on your manager's area of interest. For example, you may know how a competitor is handling an issue your company faces. You may be able to point out something to your C-level manager that they may not have heard yet from middle management.

During your conversation, keep the following points in mind.

To start the conversation, bring up something you both have in common. Be enthusiastic. Talking about the weather is often adequate "social lubrication." Of course, if you happen to know your CEO's son just got married, use that topic to open your conversation.
Bring up your main point(s).
Try to evoke a response from your audience. Leave them with a question or some interesting data. For instance, "Have you ever thought about _____?" or "Did you know that _____?" At the end of the conversation, ask for the specific thing you want, such as a meeting with that person later.

A similar brief pitch can be used when sitting next to a stranger on an airplane. First, observe the person, looking for evidence of a common ground—for example, is the person carrying luggage or a computer bag that has a special logo? Suppose you're sitting next to a person who works for a company where you'd like to work, or maybe the person is an alumnus of your university. Introduce yourself, with lots of energy and enthusiasm. Then get the person talking about what you have in common so they will be more likely to open up. Be pro-active, for example, "How would you advise me to prepare to apply for the company where you work?...What suggestions do you have for me to reach the same level as you?..." Then, at the end of your conversation, let them know that if there is any way you can help them, they should contact you. Leave your business card with them.

8.2   CREATING AND PRESENTING AN EFFECTIVE POSTER

The best posters help facilitate conversation about your work. Here are four key characteristics:

1. Impressive technical content

2. The main message appears right up front, so your viewers can understand your main idea in five seconds

3. A large enough font so your viewers can read easily. For example, if a viewer is behind the person you are talking with, or if they are in the second row of viewers, they should still be able to read your poster. Then they can formulate their questions.

4. Bullets and graphs, images, photos, symbols, and diagrams—to spark conversation

8.2.1 GUIDELINES FOR CREATING POSTERS

Start creating your poster by identifying the content you need to include. Check with the specific conference or organization to see what they expect. Here are two examples of sets of expectations for poster content. The first example is simpler and requires you to include an introduction, key concepts, a description of the problem and the results, and a conclusion that focuses on the main results.

The second example is more specific and starts with instructions for you to focus on your final deliverables or results. Here are the required parts of this example:

- Project title

- Team member names

- Names of sponsoring organization and technical liaison

- Project statement (a short version of an overview)

- Background and motivation

- Design requirements

- Design/Decision-making process

- Final deliverables

- Future implementation/Next steps

  Here are other important steps to take.

Use a storyboard to check the logical flow of your content.
When laying out your poster, be sure the eyes of your viewers know where to go. Follow the natural flow of reading English—left to right, top to bottom. Use arrows to show the flow clearly.
Make the font large enough—consider 24-point.
PowerPoint is the most common software used in creating posters, but other software for similar use includes CorelDRAW, Ink Design, Inkscape, Illustrator, LaTeX, Omnigraffle, and Photoshop.
Instead of underlining, use bold face on your poster, since it is easier to read. Also, consider blue or black text, using red to highlight, but test these colors out to see how they look. You will want to avoid colors such as yellow that disappear from three feet away.
Consider using a light background and including chunks of white between the text or graphic regions.
Use diagrams, pictures, or other graphics to supplement your text.

Print out your draft poster ahead of time in one or both of these ways:

1. Print your poster on 8 ½ by 11 paper. If you cannot read it in this size, your audience will not be able to read your poster.

2. Print your draft poster in poster size to detect other problems right away. For example, you will be able to check your colors and your graphics. In one instance, an engineer noticed the text looked fine but the pictures were garbled. They were glad they had allowed time to fix the problem.

Check with the conference or meeting ahead of time to see if one or two print shops are close by. If they are, you may not need to carry your poster with you.

Now we describe how to avoid characteristics that will make your viewers lose interest. First, avoid using too many words or more than two font families. Do not put your titles in all capital letters. Your information will be easier to understand if you use initial capital letters instead. Avoid using prose paragraphs, since no matter how large your font, long paragraphs are hard to read.

Avoid too much detail so you do not overwhelm your viewers. Do not put your entire paper on the poster, for example. In particular, do not simply lift figures, graphs, and diagrams from your paper. If you enlarge a line graph, for example, enlarge the labels on the axes and make the lines thick enough for your viewers to see.

Avoid using a gradient on your poster or on a region of your poster. If chunks of your poster are dark on the bottom and light on top, no one font color can be easily read for the entire region. Instead consider a light background, such as grey, since you will find it easy to add contrast.

Exercise 8.1. Describing your own poster.

Consider the last poster you designed. Which of the positive characteristics listed above did you use on your poster:

1.

2.

3.

4.

Which characteristics of your poster could have been improved? Explain how.

Characteristic	Ideas for Improvement
1.	
2.	
3.	

8.2.2 PREPARING TO PRESENT YOUR POSTER

The main purpose of a poster is facilitating a conversation about your work. Consider a dry run with some colleagues who are not familiar with this work. Before your practice, remember your goal will be keeping your audience engaged.

To prepare for practice:

- Prepare a very brief introduction—some suggest as short as 20 seconds. Your introduction should answer the most common first request: "Tell me about your project."

- Prepare a one- to two-minute summary as well. You may use this with people who ask for more information.

- Do not memorize your introduction or summary.

- Before practicing, brainstorm possible technical and non-technical questions your viewers may have. Think about how you might answer. You will probably be able to answer technical questions easily but you may find more general questions challenging. For instance, consider how you would answer, "Why is this problem important?"

References

For more information, refer to http://www.capstoneconf.org as one example.

Atman, Cynthia J., Sheppard, Sheri D., Turns, Jennifer, Adams, Robin S., Fleming, Lorraine N., Stephens, Reed, Streverler, Ruth A., Smith, Karl A., Miller, Ronald L., Liefer, Larry J., Yasuhara, Ken Ya, and Lund, Dennis. *Enabling Engineering Student Success: The Final Report for the Center for the Advancement of Engineering Education*. San Rafael, CA: Morgan & Claypool Publishers (2009).

Cochrane, Thomas. Enhancing the Oral-Presentation Skills of Engineering Students: Technology to the Rescue with the Virtual-I Presenter (VIP). Proceedings of 2009 American Society for Engineering Education Conference (2009).

Hattum-Janssen, Natascha Van, Adriana Fischer, and Francisco Moreira. Presentation Skills for Engineers: Systematic Intervention in a Project-Based Learning Course. *European Society of Engineering Education* (Sept. 2011).

Jamieson, Leah H., and Lohmann, Jack R. Creating a Culture for Scholarly and Systematic Innovation in Engineering Education: Ensuring U.S. engineering has the right people with the right talent for a global society. Proceedings of 2009 American Society For Engineering Education (2009).

Korte, Russell, Sheri Sheppard, and William Jordan. A Qualitative Study of the Early Work Experiences of Recent Graduates in Engineering. Proceedings of 2008 American Society for Engineering Education Conference (June 2008).

Lattuca, Lisa R, et al. *Promoting Interdisciplinary Competence in the Engineers of 2020*. Penn State: Center For the Study of Higher Education.

Lattuca, Lisa, and David Knight. In the Eye of the Beholder: Defining and Studying Interdisciplinarity in Engineering Education. Proceeding of the 2010 American Society for Engineering Education (2010).

Lattuca, Lisa R., Terenzini, Patrick T., and Volkwein, J Fredricks. *Engineering Change: A Study of the Impact of EC2000*. Center for the Study of Higher Education, Pennsylvania State University (2011).

Lappalainen, P. Integrated Language Education-a Means of Enhancing Engineers' Social Competences. *European Journal of Engineering Education*, vol. 35, no. 4 (2010), pp. 393-403. DOI: 10.1080/03043797.2010.488290

120 REFERENCES

Litzinger, Thomas A., et al. Engineering Education and the Development of Expertise. *Journal of Engineering Education*, vol. 100, no. 1 (2011), pp. 123-150. ASEE. DOI: 10.1002/j.2168-9830.2011.tb00006.x

Lohmann, Jack R., Rollins, Howard A., and Hoey, J. Joseph. Defining, developing and assessing global competence in engineers. *European Journal of Engineering Education*, vol. 31, no. 1 (March 2006), pp. 119-131. DOI: 10.1080/03043790500429906

Mitchell, Rudolph and Eng, Tony L. Assessment of Students' Learning Experience in an Oral Communication Course at MIT for EECS Majors."40th ASEE/IEEE Frontiers in Education Conference (Oct. 2010).

National Science Board. Moving Forward to Improve Engineering Education. Washington, DC: National Science Board (Nov. 2007)

Norback, Judith Shaul and Utschig, Tristan T. Assessment of Students' Responses to Workplace Oral Communication Instruction, Based on Executive Input. In preparation for submission to IEEE Transactions on Professional Communication, summer 2013.

Orr, Thomas. Assessment in Professional Communication. *IEEE Transactions on Professional Communication*, vol. 53, no.1 (2010). DOI: 10.1109/TPC.2009.2038731

Ruff, Susan, and Carter, Michael. Communication Learning Outcomes from Software Engineering Professionals: A Basis for Teaching Communication in the Engineering Curriculum. 39th ASEE/IEEE Frontiers in Education Conference (Oct. 2009).

Terenzini, Patrick T. and Lattuca, Lisa R. Benchmarking U.S. Engineering Education Vis-a-vis The Engineer of 2020. National Academy of Engineering Convocation of Professional Engineering Societies. Washington D.C. 16 May 2011. Pennsylvania State University: Center For the Study of Higher Education.

Zemliansky, Pavel, and Constance Kampf. New Landscapes in Professional Communication: The Practice and Theory of Our Field Outside the US. *IEEE Transactions on Professional Communication*, vol. 54, no. 3 (2011), pp. 221-224. DOI: 10.1109/TPC.2011.2161799

The Korte Resources, p. 110

Korte, Russell (2010). 'First, Get to Know Them': A Relational View of Organizational Socialization. *Human Resource Development International*, 13:1, 27-43. DOI: 10.1080/13678861003588984

Korte, Russell (2009). How Newcomers Learn the Social Norms of an Organization: A Case Study of the Socialization of Newly Hired Engineers. *Human Resources Development Quarterly*, 20:3, 285-306. DOI: 10.1002/hrdq.20016

Russell Korte's YouTube video on Networking: http://www.youtube.com/watch?v=3cC_EY3UWS0

About the Author

Judith Shaul Norback, Ph.D., is general faculty and Director of Workplace and Academic Communication in the Stewart School of Industrial and Systems Engineering at Georgia Institute of Technology. She applies her skills as a social psychologist to gather data from executives about stellar presentations and other oral communication skills and she conducts research on communication, designed to improve instruction. Dr. Norback has developed and provided instruction for students in industrial engineering and biomedical engineering and has advised on oral communication instruction at many other universities. The Workplace Communication Lab she founded in 2003 has had over 19,500 student visits. As of spring 2013, she shared her instructional materials with over 240 schools from the U.S., Australia, Germany, and South Korea.

Dr. Norback has studied communication and other basic skills in the workplace and developed curriculum over the past 30 years—first at Educational Testing Service, then as part of the Center for Skills Enhancement, Inc., which she founded, with clients including the U.S. Department of Labor, the National Skill Standards Board, and universities. Since 2000, when arriving at Georgia Tech, her work has focused on engineers and scientists. She has published over 20 articles in the past decade alone, including articles in *IEEE Transactions on Professional Communication*, *INFORMS Transactions on Education*, and the *International Journal of Engineering Education*.

Over the past 10 years Norback has given over 40 presentations and workshops at nation-wide conferences such as the American Society for Engineering Education (ASEE), where she currently serves as chair of her division. Norback also holds an office for the Education Forum of the Institute for Operations Research and the Management Sciences (INFORMS) and has served as Associate Chair for the Capstone Design Conference. Norback has a Bachelors' degree from Cornell University and a Masters' and Ph.D. from Princeton University. Her current research interests include increasing the reliability of the Norback/Utschig Presentation Scoring System for Engineers and Scientists and identifying the mental models students use when creating graphical representations.

Printed in the United States
by Baker & Taylor Publisher Services